AF520690

Cognitive Systems Monographs

Volume 31

Series editors

Rüdiger Dillmann, University of Karlsruhe, Karlsruhe, Germany
e-mail: ruediger.dillmann@kit.edu

Yoshihiko Nakamura, Tokyo University, Tokyo, Japan
e-mail: nakamura@ynl.t.u-tokyo.ac.jp

Stefan Schaal, University of Southern California, Los Angeles, USA
e-mail: sschaal@usc.edu

David Vernon, University of Skövde, Skövde, Sweden
e-mail: david@vernon.eu

About this Series

The Cognitive Systems Monographs (COSMOS) publish new developments and advances in the fields of cognitive systems research, rapidly and informally but with a high quality. The intent is to bridge cognitive brain science and biology with engineering disciplines. It covers all the technical contents, applications, and multidisciplinary aspects of cognitive systems, such as Bionics, System Analysis, System Modelling, System Design, Human Motion, Understanding, Human Activity Understanding, Man-Machine Interaction, Smart and Cognitive Environments, Human and Computer Vision, Neuroinformatics, Humanoids, Biologically motivated systems and artefacts Autonomous Systems, Linguistics, Sports Engineering, Computational Intelligence, Biosignal Processing, or Cognitive Materials as well as the methodologies behind them. Within the scope of the series are monographs, lecture notes, selected contributions from specialized conferences and workshops.

Advisory Board

Heinrich H. Bülthoff, MPI for Biological Cybernetics, Tübingen, Germany
Masayuki Inaba, The University of Tokyo, Japan
J.A. Scott Kelso, Florida Atlantic University, Boca Raton, FL, USA
Oussama Khatib, Stanford University, CA, USA
Yasuo Kuniyoshi, The University of Tokyo, Japan
Hiroshi G. Okuno, Kyoto University, Japan
Helge Ritter, University of Bielefeld, Germany
Giulio Sandini, University of Genova, Italy
Bruno Siciliano, University of Naples, Italy
Mark Steedman, University of Edinburgh, Scotland
Atsuo Takanishi, Waseda University, Tokyo, Japan

More information about this series at http://www.springer.com/series/8354

Manan Suri
Editor

Advances in Neuromorphic Hardware Exploiting Emerging Nanoscale Devices

Springer

Editor
Manan Suri
Department of Electrical Engineering
Indian Institute of Technology Delhi
New Delhi, Delhi
India

ISSN 1867-4925 ISSN 1867-4933 (electronic)
Cognitive Systems Monographs
ISBN 978-81-322-3701-3 ISBN 978-81-322-3703-7 (eBook)
DOI 10.1007/978-81-322-3703-7

Library of Congress Control Number: 2016958977

© Springer (India) Pvt. Ltd. 2017
This work is subject to copyright. All rights are reserved by the Publisher, whether the whole or part of the material is concerned, specifically the rights of translation, reprinting, reuse of illustrations, recitation, broadcasting, reproduction on microfilms or in any other physical way, and transmission or information storage and retrieval, electronic adaptation, computer software, or by similar or dissimilar methodology now known or hereafter developed.
The use of general descriptive names, registered names, trademarks, service marks, etc. in this publication does not imply, even in the absence of a specific statement, that such names are exempt from the relevant protective laws and regulations and therefore free for general use.
The publisher, the authors and the editors are safe to assume that the advice and information in this book are believed to be true and accurate at the date of publication. Neither the publisher nor the authors or the editors give a warranty, express or implied, with respect to the material contained herein or for any errors or omissions that may have been made.

Printed on acid-free paper

This Springer imprint is published by Springer Nature
The registered company is Springer (India) Pvt. Ltd.
The registered company address is: 7th Floor, Vijaya Building, 17 Barakhamba Road, New Delhi 110 001, India

To Geeta, Pramod, Aarushi, and Geetika

कर्मण्यकर्म यः पश्येदकर्मणि च कर्म यः ।
स बुद्धिमान्मनुष्येषु स युक्तः कृत्स्नकर्मकृत् ॥ १८ ॥

Shrimad Bhagwad Gita [Chapter 4: Verse 18]

**" He who sees inaction in action, and action in inaction
Is spiritually wise, transcendentally situated a perfect performer of all actions "**

Preface

The advent of cheap electronic sensors, cloud computing, IoT, smart devices, mobile computing platforms, driverless cars, drones, etc., has led to generation of enormous amounts of data. Some characteristics central to this big data are its asynchronous and non-standardized nature. The vast amount of data by itself is of less value; however, the ability to effectively and efficiently process it in real-time leading to meaningful patterns, trends, and interpretation is the real treasure trove.

Several upcoming unconventional (non-Von Neumann) computing paradigms, where memory (storage) and processing are not isolated tasks in themselves or rather *memory is intelligent*, offer promising capabilities to this problem of massive non-synchronous, non-standardized data treatment. Techniques such as software artificial neural networks (ANNs), artificial intelligence (AI), and machine learning (ML) have been proving their mettle in fields as diverse as autonomous navigation, to robotics to analytics since a while. However the full potential of these computing paradigms can only be realized when they are directly implemented on dedicated low-power, compact, reconfigurable, programming-free hardware.

When it comes to dedicated hardware, some first contenders are CMOS-ASICs, DSPs, GPUs, and FPGAs. However, most of these implementations rely on a layer of digital (Von Neumann modified) abstraction even if some grassroots computing arises out of purely analog traits.

To this end, over the last few years there has been a lot of activity across research groups postulating efficient hybrid CMOS-"*nanodevice*" computing hardware architectures. The "*nanodevice*" in these hybrid systems cover a vast range of technologies such as organic nanoparticle transistors (NOMFETs), carbon nanotubes (CNTs), atomic nanogap switches, silicon thin-film transistors (TFTs), magnetic spin-based devices, to families of emerging non-volatile resistive memory including phase-change memory (PCM), conductive bridge memory (CBRAM or PMC), metal-oxide-based memory (OxRAM), theoretical memristor, and so on, to name a few.

This book is a selective collection of recent works from some of the leading research groups across the globe working to achieve dedicated hybrid (CMOS + *nanodevice*) hardware for neuromorphic computing. The book in its present form is certainly not exhaustive of all the research in the field, but is an attempt to get started; bringing valuable and diverse approaches on the subject matter under one roof.

While curating the valuable contributions for the present edition, special attention was paid to create a right mix of conceptual (primarily simulation-based studies) and experimental (technology-based studies) works.

The book may be used as teaching material for undergraduate and postgraduate course work, or a focused read for advanced researchers working in the domain and related areas.

Key building blocks for neuromorphic systems would comprise of hardware implementations of—(i) the *neuron* or *neuron-like* functionality and (ii) the *synapse* or *synapse-like* functionality, where both are powered by relevant *learning rules*.

Chapter "Hardware Spiking Artificial Neurons, Their Response Function, and Noises," by researchers at *Korea Institute of Science and Technology*, provides an overview of silicon hardware implementation of the first essential block of neuromorphic systems, i.e., the *neuron*.

Chapter "Synaptic Plasticity with Memristive Nanodevices," by researchers at *University of Lille—France*, offers the reader a strong overview or primer on different techniques of emulating forms of synaptic plasticity using various memristive nanodevices.

Chapter "Neuromemristive Systems: A Circuit Design Perspective," by researchers at *University of Rochester—USA*, focuses on hybrid circuit design perspectives of emerging neuromemristive architectures and systems.

Chapter "Memristor-Based Platforms: A Comparison Between Continous-Time and Discrete-Time Cellular Neural Networks," by researchers at *Politecnico di Torino—Italy*, focuses on analysis and comparison of memristive continuous-time and discrete-time cellular neural networks.

Chapter "Reinterpretation of Magnetic Tunnel Junctions as Stochastic Memristive Devices," by researchers at *University Paris-Sud—France*, discusses how spin-transfer torque magnetic random access memory (STT-MRAM) can be used to realize stochastic neuroinspired hardware architectures.

Chapter "Multiple Binary OxRAMs as Synapses for Convolutional Neural Networks," by researchers at *French Alternative Energies and Atomic Energy Commission* (*CEA-LETI and CEA-LIST*), presents the implementation of convolutional neural networks exploiting metal-oxide-based OxRAM technology.

Chapter "Nonvolatile Memory Crossbar Arrays for Non-von Neumann Computing," which is a joint effort of researchers at *EPFL—Switzerland, IBM Almaden—USA, and Pohang University—Korea*, presents neuromorphic implementations that utilize chalcogenide-basedphase-change memory (PCM), and non-filamentary RRAM (PCMO)-based nanodevices.

Chapter "Novel Biomimetic Si Devices for Neuromorphic Computing Architecture," by researchers at *Indian Institute of Technology Bombay*, presents novel SiGe-based nanodevices for neuron and synaptic implementations inside the neuromorphic hardware architectures.

Chapter "Exploiting Variability in Resistive Memory Devices for Cognitive Systems," by researchers at *Indian Institute of Technology—Delhi*, presents implementation of extreme learning machines (ELM) and restricted Boltzmann machines (RBM) using RRAM nanodevices.

Chapter "Theoretical Analysis of Spike-Timing-Dependent Plasticity Learning with Memristive Devices," by researchers at *University Paris-Sud—France*, and *Alternative Energies and Atomic Energy Commission* (*CEA-LIST*), presents the underlying theoretical framework behind STDP-based learning and its equiv

New Delhi, India Manan Suri

Contents

Hardware Spiking Artificial Neurons, Their Response Function, and Noises 1
Doo Seok Jeong

Synaptic Plasticity with Memristive Nanodevices 17
Selina La Barbera and Fabien Alibart

Neuromemristive Systems: A Circuit Design Perspective 45
Cory Merkel and Dhireesha Kudithipudi

Memristor-Based Platforms: A Comparison Between Continous-Time and Discrete-Time Cellular Neural Networks 65
Young-Su Kim, Sang-Hak Shin, Jacopo Secco, Keyong-Sik Min and Fernando Corinto

Reinterpretation of Magnetic Tunnel Junctions as Stochastic Memristive Devices 81
Adrien F. Vincent, Nicolas Locatelli and Damien Querlioz

Multiple Binary OxRAMs as Synapses for Convolutional Neural Networks 109
E. Vianello, D. Garbin, O. Bichler, G. Piccolboni, G. Molas, B. De Salvo and L. Perniola

Nonvolatile Memory Crossbar Arrays for Non-von Neumann Computing 129
Severin Sidler, Jun-Woo Jang, Geoffrey W. Burr, Robert M. Shelby, Irem Boybat, Carmelo di Nolfo, Pritish Narayanan, Kumar Virwani and Hyunsang Hwang

Novel Biomimetic Si Devices for Neuromorphic Computing Architecture 151
U. Ganguly and Bipin Rajendran

Exploiting Variability in Resistive Memory Devices for Cognitive Systems 175
Vivek Parmar and Manan Suri

Theoretical Analysis of Spike-Timing-Dependent Plasticity Learning with Memristive Devices 197
Damien Querlioz, Olivier Bichler, Adrien F. Vincent and Christian Gamrat

Erratum to: Novel Biomimetic Si Devices for Neuromorphic Computing Architecture E1
U. Ganguly and Bipin Rajendran

About the Editor

Dr. Manan Suri (Member, IEEE) is an Assistant Professor with the Department of Electrical Engineering, Indian Institute of Technology-Delhi (IIT-Delhi). He was born in India in 1987. He received his PhD in Nanoelectronics and Nanotechnology from Institut Polytechnique de Grenoble (INPG), France in 2013. He obtained his M.Eng. (2010) and B.S (2009) in Electrical & Computer Engineering from Cornell University, USA. Prior to joining IIT-Delhi in 2014, he has worked as a part of the Central R&D team, NXP Semiconductors, Belgium and Advanced Non-Volatile Memory Technology Group, CEA-LETI, France. His research interests include Non-Volatile Memory Technology, Unconventional Computing (Machine-Learning/Neuromorphic), and Semiconductor Devices. He holds several granted and filed US, and international patents. He has authored books/book chapters and 30+ papers in reputed international conferences and journals. He serves as committee member and reviewer for IEEE journals/conferences. He is a recipient of several prestigious national and international honors such as the IEI Young Engineers Award, Kusuma Outstanding Young Faculty Fellowship, and Laureat du Prix (NSF-France).

Hardware Spiking Artificial Neurons, Their Response Function, and Noises

Doo Seok Jeong

Abstract In this chapter, overviewed are hardware-based spiking artificial neurons that code neuronal information by means of action potential, *viz.* spike, in hardware artificial neural networks (ANNs). Ongoing attempts to realize neuronal behaviours on Si 'to a limited extent' are addressed in comparison with biological neurons. Note that 'to a limited extent' in this context implicitly means 'sufficiently' for realizing key features of neurons as information processors. This ambiguous definition is perhaps open to a question as to what neuronal behaviours the key features encompass. The key features are delimited within the framework of neuromorphic engineering, and thus, they approximately are (i) integrate-and-fire; (ii) neuronal response function, i.e. spike-firing rate change upon synaptic current; and (iii) noise in neuronal response function. Hardware-based spiking artificial neurons are aimed to achieve these goals that are ambitious albeit challenging. Overviewing a number of attempts having made up to now illustrates approximately two seemingly different approaches to the goal: a mainstream approach with conventional active circuit elements, e.g. complementary metal-oxide-semiconductor (CMOS), and an emerging one with monostable resistive switching devices, i.e. threshold switches. This chapter will cover these approaches with particular emphasis on the latter. For instance, available types of threshold switches, which are classified upon underlying physics will be dealt with in detail.

1 Introduction

Neuromorphic systems based on full hardware artificial neural networks consist of a complex array of building blocks encompassing artificial neuron and synapse devices. The artificial neuron represents the information—input synaptic current—which it receives by relaying the corresponding response to neighbouring neurons via

D.S. Jeong (✉)
Center for Electronic Materials Research, Korea Institute of Science and Technology, Hwarangno 14-gil 5, Seongbuk-gu, Seoul 136-791, Republic of Korea
e-mail: dsjeong@kist.re.kr

© Springer (India) Pvt. Ltd. 2017

M. Suri (ed.), *Advances in Neuromorphic Hardware Exploiting Emerging Nanoscale Devices*, Cognitive Systems Monographs 31,
DOI 10.1007/978-81-322-3703-7_1

synapses. Namely, the neuron encodes the input information into a particular form of response and the response is subsequently processed by the synapses and further encoded by the bridged neurons. Encoded information at the outset, for instance, by sensory neurons flows throughout the entire network by repeating the aforementioned process. The neuron is therefore required to represent information clearly in order to prevent information loss in the network. Hereafter, a device in charge of information representation in a neuromorphic system is referred to as an artificial neuron—in short, neuron—for simplicity, at times its behaviour is barely biologically plausible though.

Choice of type of an artificial neuron depends upon the dimension of a neuromorphic system—including time frame, i.e. dynamic system, leads to an additional request for dynamic neuronal behaviour—and the information type in use such as binary, multinary, and analogue. The simplest case is when the neuromorphic system disregards time frame, i.e. static, and employs information in binary; time-independent binary neurons that merely represent 1 bit of information meet the requirements. For instance, a summing amplifier—summing inputs from adjacent neurons—in conjunction with a single transistor, representing binary states—channel on and off ($\log_2 n$ and $n = 2$, where n denotes the number of states)—upon the input in total, perhaps successfully works as an artificial neuron in this simplest system. Employing multinary or analogue information in the neuromorphic system requires the neuron to exhibit various states ($n > 2$) for $\log_2 n$ bit of information; for instance, a transistor, working in the subthreshold regime, and thus representing multiple states of channel conductance, may meet the need. Note that the use of multinary information gains the remarkable benefit that the same amount of information is represented by much less number of multinary neurons than binary neurons if preventing the response variability upon the same input and the consequent information loss [1, 2]. Later, this variability-induced information loss will be addressed in detail from the perspective of information decoding.

When it comes to time-dependent (dynamic) neuromorphic systems, the neuron should introduce a time-varying response to the input in total, which also varies in time, rendering the neuron complicated. If binary, the input elicits all—or—nothing output. The rest case, dynamic and multinary information-utilizing neuromorphic systems, is most biologically plausible regarding information representation by biological neurons. As for the dynamic binary system, the response of the neuron should be reliably distinguished in time. Besides, the response should vary upon the input in order for the response to be decodable. For instance, in case of a single-pulse output, the response may be parameterized by pulse height and/or width—varying with respect to the input—and used in information representation. A number of different types of neurons and their outputs can be used in such a system as far as they meet the aforementioned requirements and are compatible with the synapse as a whole in the system. Regarding the root of neuromorphic engineering, our intuition perhaps leads us to use the most biologically plausible one—satisfying fidelity to key features of biological neurons—among the possible candidates. Thus far, a great deal of effort on building biologically plausible artificial neurons—hereafter such a type of a neuron is referred to as *spiking* neuron—

has been made by capturing key features of biological neurons and then realizing them by means of electric circuit components [3].

The forthcoming Sects. 1.1–1.3 are dedicated to addressing essential characteristics of biological neurons, which artificial neurons are required to reflect. Sect. 1.4 is dedicated to introducing in silico—computational—neuron models that also form the basis of hardware artificial neurons. Section 2 addresses the state-of-the-art spiking artificial neurons being classified as silicon neuron (SiN), realized by very-large-scale integration (VLSI) circuits, and functional materials-based emerging technologies.

1.1 Biological Neurons

Biological neurons (also known as nerve cells) are a *living* electrochemical system being made of a lipid membrane containing ion channels and pumps. A schematic of a biological neuron is illustrated in Fig. 1. The membrane separates intracellular and extracellular media in which there exist significant differences in concentration of important ions, e.g. Na^+, K^+, Ca^{2+}, and Cl^-, between the media. Na^+, Cl^-, and Ca^{2+} are rich in the extracellular medium whereas K^+ and fixed anions—in the sense that they cannot diffuse out through ion channels—such as organic acids and proteins, are rich on the other side. Namely, for each ion, chemical potential through the membrane is not equal. In the resting state, the extracellular charge in total is positive, whereas the total charge in the intracellular medium is negative as a whole; that is, the resting state indicates electrical polarization, leading to electric field, e.g. electrostatic potential gradient, evolution through the lipid membrane. This potential difference caused by the difference in chemical potential between the two sides of the membrane is referred to as the Donnan potential [4], which tends to recover the equilibrium distribution—no chemical potential gradient—of each ion. This ionic distribution is not spontaneous configuration as being far from energy-minimizing conditions; work should be done by a third party to maintain the polarized state, raising the internal energy. Ion pumps embedded in the membrane

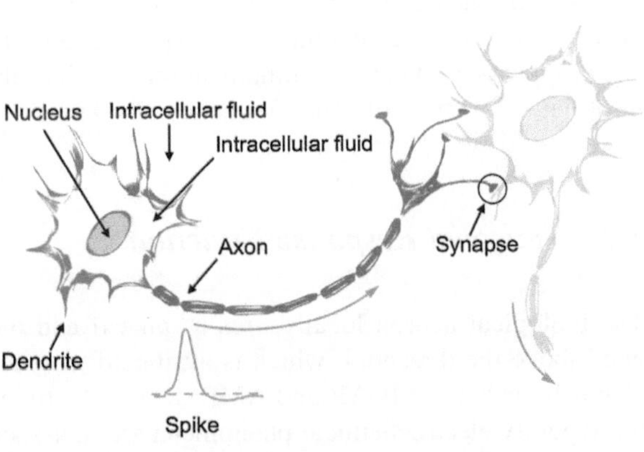

Fig. 1 Schematic of neurons and chemical synapses in-between. The *grey* and the *orange* neurons denote a presynaptic and a postsynaptic neurons, respectively. The *blue arrows* indicate the spike propagating direction—from the pre- to the postsynaptic neuron

are in charge of the work by pumping out ions in order for the resting state to be maintained [5]. Ion pumps consume chemical energy that is subsequently converted into electrical energy as for batteries [6]. They are known as sodium–potassium adenosine triphosphatase (Na^+/K^+–ATPase), which drive Na^+ and K^+ ions—two important ions in membrane potential maintenance—into and out of the intracellular fluid, respectively.

Temporary breakdown of the resting state occurs upon significant fluctuation in ion concentration in the intracellular fluid. In particular, Upon release of chemical messengers, i.e. neurotransmitters, at a chemical synapse, N-methyl-D-aspartate receptor (NMDAR) and α-amino-3-hydroxy-5-methyl-4-isoxazolepropionic receptor (AMPAR) open their ion channels, so that both monovalent ions, e.g. Na^+ and K^+, and divalent ions, e.g. Ca^{2+}, are able to diffuse into the intracellular fluid, leading to breakdown of the resting state ion concentration [7]. The inward diffusion corresponds to *synaptic current* that is of significant importance in spiking. Such concentration fluctuation causes a change in membrane potential regarding the aforementioned origin of the potential. Given that Na^+ and K^+ ion channels are voltage-gated [5], the change in the membrane potential can determine the channel conductance. However, the consequent change in the membrane potential should be larger than a *threshold* for the channel opening; otherwise, the change lasts for a short time without leading to any significant phenomena. Thus, introduction of synaptic current—sufficing for large potential evolution—opens voltage-gated Na^+ and K^+ channels, and thus, the resting state ion distribution is locally destroyed by inward Na^+ ions and outward K^+ ions through their characteristic ion channels. As a consequence, the membrane undergoes depolarization—down to almost zero volt or sometimes polarity reversal occurs.

Once such local depolarization occurs, ion pumps do not let the membrane maintain the depolarization and recovers the resting state ion distribution, and thus membrane potential. Meanwhile, this local depolarization in turn affects adjacent Na^+ and K^+ ion channels and the same depolarization procedure keeps being repeated throughout the entire axon down to the end of the axon that is termed as axon terminal—similar to dominoes [6]. Therefore, so far explained procedure in both time and spatial frames produces an action potential also known as spike and it propagates along the axon in due course. Neurons in a neural network communicate by firing spikes and receiving them. To be precise, a train of spikes, rather than a single spike, appears used in the communication, which enables representation of analogue information in terms of spike frequency also known as activity.

1.2 Neuronal Response Function

The biological neuron locally fires a spike if and only if the membrane potential goes above the threshold, which is attributed to synaptic current injection through channels such as NMDAR and AMPAR. For the time being, the question as to how these purely electrochemical phenomena are in association with information representation arises. To answer this question, we need to elucidate the electrochemical phenomena in regard to a relationship between input and output and define types of both analogue input and output. Given that synaptic current injection initiates spiking at the outset, synaptic current works as input, varying its quantity upon synaptic weight. Synapse in this chapter implies chemical synapse unless otherwise stated. Synapses are spatially discrete given that each neuron has an immense number of synapses. Also, the synaptic current through them varies in time upon arrival of a train(s) of spikes from the presynaptic neuron. Fortunately, the membrane of the neuron resembles a capacitor that integrates the synaptic current varying in metric as well as temporal space. The integrated synaptic current at given time is therefore typically taken as input value. Activity—the number of spikes per unit time or at times per particular time period—caused by the input is typically regarded as output; that is, a train of spikes, rather than a single spike, is of concern in determining output value. Note that evaluating time-varying activity is fairly complicated, in particular, when the synaptic current varies in time—this situation is very often encountered in in vivo experiments. There exist several methods in the linear-filtering framework with different linear filters (see Ref. [8]).

The activity tends to increase with synaptic current and this tendency is the substrate of the neuronal response function that enables representation of analogue information. Given the role of the membrane as a capacitor, the larger the synaptic current, the sooner the membrane potential hits the threshold for spiking and the more there exist spikes in a unit time. This tendency holds for stereotypical neurons, the detailed behaviour is rather complicated in the light of presence of refractory time though. A relation between activity and synaptic current is referred to as a gain function, providing the key features of the tuning function [9]. From a perspective of information coding, the gain function provides single neuron's ability to encode a large number of different inputs—in comparison with digital components such as a transistor representing merely 1 bit of information—and make them distinguishable by the output, i.e. very clearly decodable in one-to-one correspondence. However, biological neurons are noisy [10–12], in particular, in vivo neurons, so that a great deal of information is lost. Neuronal noises and related information loss will be addressed in the following subsection in detail.

1.3 Neuronal Noises

Neuronal behaviour, particularly in vivo behaviour, is noisy including a number of unpredictable features. Neuronal noises are loosely classified as (i) background noise, e.g. white noise, in the membrane potential in conjunction with the neuronal response of focus [8] and (ii) irregularity in spiking, e.g. irregular periodicity of spikes in a single spike train [1, 2, 13, 14]. The former may be attributed to white-noise-like synaptic current injection that keeps on perturbing the membrane potential. [13, 15] These two types of noises may be correlated insomuch as the random synaptic current injection—causing the former—enables random spike firing at times [15]. Additionally, the background noise is thought to enable gain modulation, altering the gain function [16]. For the time being, questions as to what roles of the noise in a neural network are and if it is good or bad have barely been answered clearly. Regarding information representation, of course, the noise is the root cause of information loss, and thus, it needs to be avoided [17]. The brain may, however, perform stochastic calculations making use of probability, rather than single bit, coding schemes, which can be robust to errors due to noises in the noisy neural network [18–20]. However, this hypothesis is open to objection due to lack of understanding of probability estimation [14].

Regarding the irregularity in spiking, the irregularity is well parameterized by variability in inter-spike interval (ISI) in an ensemble of spike trains [8]. In the biological neuron, the ISI distribution in the same spike train is dispersive. The distribution varies upon experimental conditions such as in vitro and in vivo experiments. Despite such variations, it is generally agreed that the biological neuron represents Poisson or Poisson-like noises, which is possibly justified by the relation between the mean neuronal response and the corresponding variance [8, 13]. The Poisson noise is a result of random spike generation, so that it is also required to be identified, for instance, by simply evaluating correlation between the spikes in the same train [21]. A similar type of noise may be present in hardware artificial neurons due to operational variability in the core device. We will revisit this later in Sect. 2.

1.4 Artificial Neuron Models

Thus far, biological neurons are featured by (i) integrate-and-fire—the membrane integrates synaptic current and fires a spike when the resulting membrane potential goes above the threshold; (ii) gain function—the output varies upon the input and they are in one-to-one correspondence in the ideal case; and (iii) noise leading to variability in ISI, and thus activity. There exist several artificial neuron models, being mostly for computational purpose but serving as the basis of hardware-based artificial neurons, which mainly reflect the first two features. The integrate-and-fire (IF) model is simplest; its equivalent circuit is shown in Fig. 2a. The capacitor in Fig. 2a corresponds to the membrane of a biological neuron, and it integrates the injected synaptic current until the electrostatic potential difference across the capacitor reaches the threshold. Crossing the threshold generates a spike by the lumped switch in parallel with the capacitor, and the refractory period follows spiking hindering another spiking in close succession. The lumped switch and its functions in spiking can simply be programmed, rendering the model very easily implementable. However, deficient fidelity of the model to the biological neuron is noticeable at a glance; particularly, the membrane cannot be lossless perfect dielectric as assumed in the model in the light of leakage of ions through the membrane. That is, underpinning the model and thus making it more realistic require a parallel resistor to the capacitor, which results in charge loss in due course when the model system is subject to no or smaller synaptic current than it used to be. This charge loss and

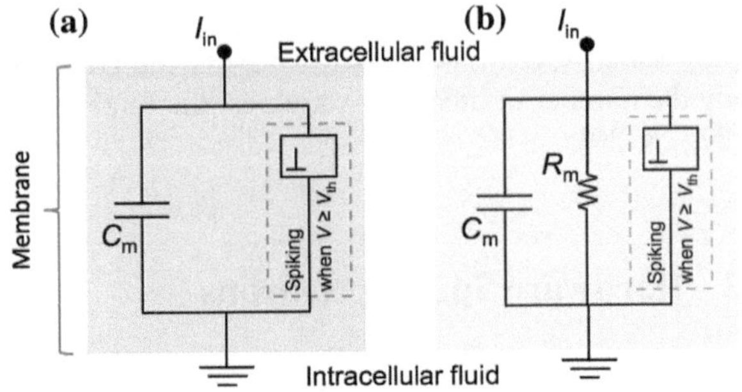

Fig. 2 Equivalent circuits of (**a**) IF and (**b**) LIF neurons. For each circuit, the component lumped in the box outlined with a *grey dashed line* generates a spike when the membrane potential crosses V_{th}, i.e. a threshold membrane potential for spiking

the corresponding potential decay away with time cannot be implemented in the IF model. This modified model is referred to as the leaky integrate-and-fire (LIF) model [8, 17, 22–25]. The equivalent circuit is illustrated in Fig. 2b. The membrane potential V_m with time in the subthreshold regime is described by the following equation: $V_m = I_{in} R_m \left(1 - e^{-t/\tau}\right)$, where I_{in}, R_m, and τ mean the input current, i.e. synaptic current, the membrane resistance, and the time constant for membrane charging and discharging—product of R_m and C_m (membrane capacitance)—respectively. Including the parallel resistor barely adds up further complexity so that the LIF model is still easy to implement in in silico neural networks. In fact, the LIF model is thought to reflect the most essential feature in a very simple manner, offering remarkable computational efficiency regarding computational time. In addition, this model likely bridges a computational neuron model to a hardware-achievable model in which realizing the lumped switch by means of scalable electrical components is of significant concern. This issue will be revisited in Sect. 2.

It should be noted that the resting state of both IF and LIF models indicates depolarization and, upon incidence of synaptic current, a build-up of membrane potential, i.e. polarization, evolves in due course unlike the biological neuron—membrane polarization in the resting state and depolarization evolving upon synaptic current incidence. It is, however, believed that this reversal is trivial. The most significant discrepancy between the simple LIF model and the biological neuron consists in the fact that the LIF deals with a single type of charge, whereas the biological neuron involves two different ions, i.e. Na^+ and K^+, and their transport through different ion channels in the membrane. In this regard, the LIF model is often regarded as being oversimplified to gain computational efficiency.

Another class of artificial neuron models is conductance-based LIF model in which the conductance of the ion channel varies upon arrival of a spike from the postsynaptic neuron [26–29]. This class is more biologically plausible in the sense that the assumed channel conductance increase upon spike transmission is the feature of the biological neuron albeit lumped. Furthermore, unlike the integrate-and-fire-based models, the resting state denotes polarization and the polarization is perturbed by the increase in the channel conductance, leading to depolarization. Various conductance-based models are available, ranging from a simple one offering similar computational efficiency to the LIF model [29] to a one, offering a great deal of fidelity to the biological neuron, such as the Hodgkin–Huxley (HH) model [26]. Later, several attempts to simplify the HH model have been made by reducing the number of auxiliary variables; the FitzHugh–Nagumo model is an example [27, 28, 30].

2 Hardware Spiking Neurons

Several prototypical hardware spiking neurons have been achieved in search of models fulfilling the aforementioned requirements for neuronal information representation. Basically, they should produce spike trains in association with input synaptic current and the activity of the neuronal response is required to vary upon the input synaptic current. When it comes to hardware neurons, integration, for which scalability is a premise, should be taken into account, and thus scalability is an additional requirement.

2.1 Silicon Neurons

The neuromorphic engineering stemmed from VLSI technologies, dating back in 1980s [31]. In the past few decades, vigorous attempts to realize neuromorphic circuits, encompassing basic building blocks, i.e. neuron and synapse, and related physiological phenomena, have been made by utilizing VLSI circuit elements such as field-effect transistor (FET), capacitor, and resistor [3, 32, 33]. As a consequence, they have led the mainstream trend of neuromorphic engineering. An evident advantage of the SiN is such that designed circuits can readily be fabricated using the conventional complementary metal-oxide-semiconductor (CMOS) technology. Also, the SiN is likely scalable unless the case of real-time operation, i.e. the operational timescale is comparable to that of biological neurons, which perhaps requires a large capacitor(s) working as the membrane. Note that the same difficulty in scalability due to a large capacitor also holds for the emerging neurons. A number of SiNs mimicking different in silico neuron models and the progress in the related technologies are very well overviewed by Indiveri et al. in their review paper [3].

2.2 Emerging Spiking Neurons

As mentioned in Sect. 1.4, the LIF model includes the lumped switch and one should be in search of the switch—achievable by means of electrical components—so as to realize the LIF model. In 2012, Pickett et al. at the Hewlett Packard Labs proposed a neuron model in the framework of the LIF neuron [34]. Later, Lim et al. named the model the neuristor-based LIF (NLIF) model [21]. In the NLIF circuit, a neuristor—

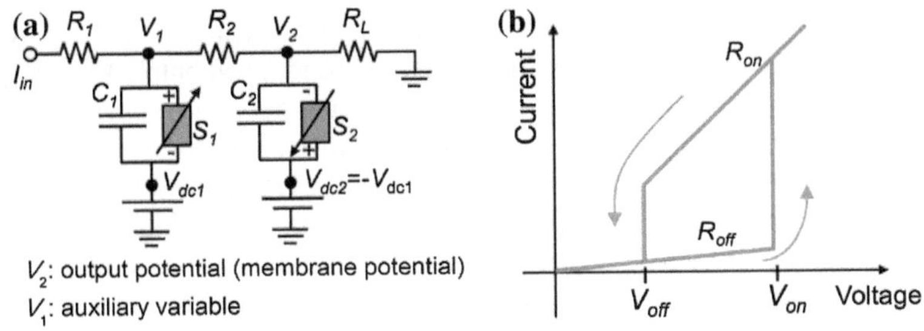

Fig. 3 **a** Equivalent circuit of the NLIF neuron model. The *blue boxes* stand for the threshold switch whose current–voltage hysteresis loop is sketched in **b**. Reprinted with permission from Ref. [21]. Copyright 2015, Nature Publishing Group

its first proposal dates back to 1962 [35]—is employed as the lumped switching; the neuristor consists of a pair of Pearson–Anson oscillators in parallel [34]. The oscillators are made of a pair of a capacitor and a *threshold switch* in parallel. The circuit of the proposed NLIF neuron circuit is illustrated in Fig. 3a. The key role in the oscillation is played by the S-shape negative differential resistance (NDR) effect that is not followed by memory effect. This class of resistive switching is termed as 'monostable' resistive switching in order to distinguish it from 'bistable' switching typically referring to memory-type resistive switching [36]. Given the lack of memory effect, this class of switching is also referred to as volatile switching. The threshold switching effect is a typical example of volatile switching. The effect has been seen in various systems such as amorphous higher chalcogenides [37–40], Mott insulators [34, 36, 41, 42], Si n+/p/n+ junctions [43], and particular transition metal oxides such as NbO_x [44, 45].

A typical current–voltage (I-V) hysteretic loop of a threshold switch is shown in Fig. 3b. Two thresholds for critical resistance changes are defined in Fig. 3b: V_{on} for high-to-low resistance transition and V_{off} for the reversed one. The low resistance state (LRS; R_{on}) emerges at V_{on}, and a drastic transition from the high resistance state (HRS; R_{off}) to the LRS takes place. The LRS can be maintained unless the applied voltage falls below V_{off} ($V_{off} < V_{on}$), i.e. the HRS is of the only stability under no voltage application.

Regarding the working principle of the NLIF model, the membrane potential evolution and the resulting spiking can be described by membrane potential V_2 and auxiliary variable V_1 as follows [21]:

$$C_1 \frac{dV_1}{dt} = I_{in} - \frac{1}{R_{S1}}(V_1 - V_{dc1}) - \frac{1}{R_2}(V_1 - V_2) \tag{1}$$

and

$$C_2 \frac{dV_2}{dt} = \frac{1}{R_2}(V_1 - V_2) - \frac{1}{R_{S2}}(V_2 - V_{dc2}) - \frac{1}{R_L} V_2, \tag{2}$$

where R_{S1} and R_{S2} denote the resistance of threshold switches S1 and S2, respectively. The spiking dynamics of the NLIF model can be mapped onto a $V_2 - V_1$ phase plane; this phase-plane analysis clearly shows the dynamics in a compact manner. In fact, such a phase-plane analysis is often employed to account for membrane potential change in time for the FitzHugh–Nagumo model [27, 28]. Towards this end, V_1- and V_2-nullclines need to be defined, which denote the lines on which every (V_2, V_1) point satisfies $\mathrm{d}V_1/\mathrm{d}t$ and $\mathrm{d}V_2/\mathrm{d}t$, respectively. The V_1- and V_2-nullclines are readily obtained by equating the left-hand side of Eqs. (1) and (2) to zero as follows:

$$V_1 - \frac{R_{S1}}{R_2 + R_{S1}} V_2 - \frac{R_2 V_{dc1}}{R_2 + R_{S1}} - \frac{R_2 R_{S1}}{R_2 + R_{S1}} I_{in} = 0, \tag{3}$$

and

$$V_1 - R_2 \left(\frac{1}{R_2} + \frac{1}{R_{S2}} + \frac{1}{R_L} \right) + \frac{R_2 V_{dc2}}{R_{S2}} = 0, \tag{4}$$

respectively.

A crossing point between these nullclines is termed as fixed point at which both $\mathrm{d}V_1/\mathrm{d}t$ and $\mathrm{d}V_2/\mathrm{d}t$ are zero—no gradient in either axis exists so that the (V_2, V_1) configuration becomes stuck at this stable point unless perturbation is applied to the configuration, for instance, by changing I_{in}. It should be noted that, as explained, R_{S1} and R_{S2} vary upon the threshold switching defined by two thresholds, V_{on} and V_{off}, so that the nullclines also vary in association with R_{S1} and R_{S2} as shown in Eqs. (3) and (4). This change in the nullclines features the spiking dynamics of the NLIF model, rendering this model distinguishable from other dynamic models such as the HH and the FHN models [21]. The results of the phase-plane analysis under a constant synaptic current are plotted in Fig. 4, sorted in due course. Following this first cycle of (V_2, V_1) at the outset, the trajectory repeats the grey cycle plotted in Fig. 4f, which is referred to as limit cycle.

The limit cycle shows important information as to the spiking dynamics at a glance: V_2 maximum (spike height) and V_2 minimum (spike undershoot). In addition, such phase-plane analysis intuitively provides a current threshold for spiking in comparison between Figs. 4a, b—the fixed point without synaptic current application (Fig. 4a) is required to be driven out of the subthreshold region in order for spiking to be initiated as for the case shown in Fig. 4b. The fixed point for the nullclines in Fig. 4b is definitely placed in the above-threshold region, so that the trajectory meets the threshold switching condition. Otherwise, a fixed point stays in the subthreshold region and thus no threshold switching takes place throughout the entire current application.

A following question is if the NLIF model successfully represents a gain function. Recent calculations on the NLIF model have properly answered the question by exhibiting an *ideal* gain function shown in Fig. 5a. The activity tends to increase with synaptic current above the threshold for spiking. In particular, for case α, the change in activity with respect to synaptic current is very large compared to the other cases—suitable for discriminating the applied synaptic current without uncertainty [21].

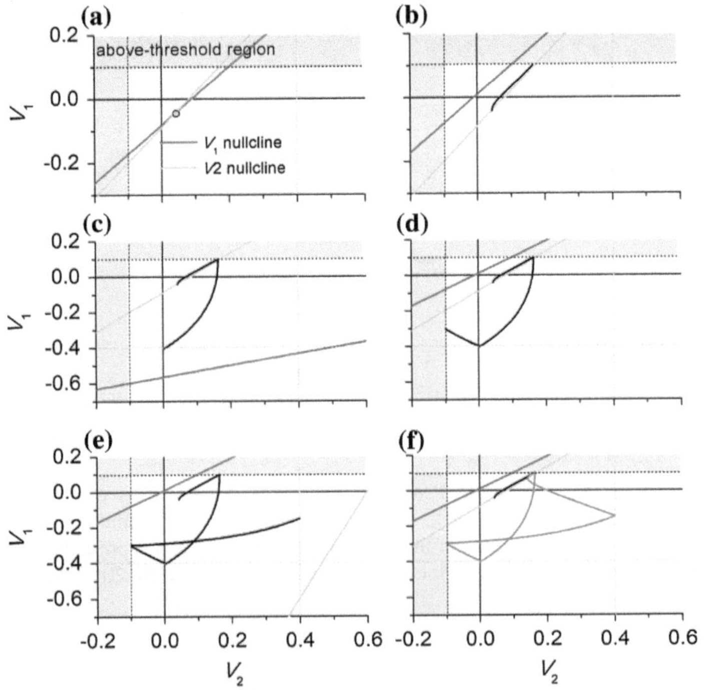

Fig. 4 Phase-plane analysis on the NLIF neuronal dynamics. The V_1- and V_2-nullclines Eqs. (3) and (4) cross each other at a particular point—fixed point—on the plane. The resting state is described by the nullclines and fixed point in **a**. The state of the neuron stays at the fixed point unless perturbation. **b** Upon application of a current, the V_1-nullclile shifts upwards and the fixed point leaves the subthreshold region; the trajectory moves towards the fixed point in due course. **c** The trajectory confronts the threshold switching condition, so that V_1-nullclile shifts down according to Eq. (3) and accordingly a new fixed point appears. **d–f** Threshold switching of both switches subsequently takes place and the dynamic procedure finishes a single complete cycle, i.e. limit cycle,—a *grey* cycle in **f**. The cycle in succession follows this limit cycle and the same dynamics is repeated. Reproduced with permission from Ref. [21]. Copyright 2015, Nature Publishing Group

Owing to the gain function, the tuning function of the NLIF model is consequently acquired by presuming the synaptic current in association with a stimulus and the preferred stimulus of the synapse [21]. Invoking the Bienenstock–Cooper–Munro rule, stimulus selectivity is spontaneously given to synapses so that each synapse is supposed to possess strong preference for a unique stimulus [46]. In case of one-dimensional stimulation, such as edge orientation and light intensity in visual stimulation, the synaptic current can be assumed to be of a bell-shaped, i.e. Gaussian, function centred at the preferred stimulus (zero degree; see Fig. 5b), which eventually gives a tuning function (see Fig. 5c). The tuning function is also bell shaped and represents the maximum activity at the preferred stimulus. One of the most significant implications of a tuning function is to bridge external sensory information, i.e. stimulus, to the corresponding neuronal response. In this way, the sensory

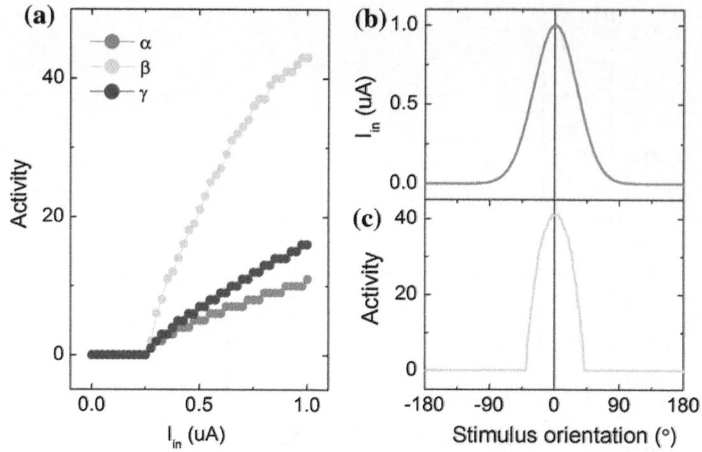

Fig. 5 **a** Gain functions of the NLIF neuron with three different sets of circuit parameters. **b** Assumed synaptic current profile with respect to one-dimensional stimulus (here, bar orientation). **c** The resulting tuning function of the given NLIF neuron clearly represents good selectivity to input stimulus. Reproduced with permission from Ref. [21]. Copyright 2015, Nature Publishing Group

information is encoded by the neuron into the value for activity and later the activity is most likely decoded in communication between adjacent neurons.

It is well known that resistive switching encompasses variability in switching parameters such as V_{on}, V_{off}, R_{on}, and R_{off}. The same most likely holds for the threshold switches in the NLIF model. The variation of such switching parameters is generally a switching event-driven, rather than real-time-varying, phenomenon, i.e. upon termination of one switching cycle—depicted in Fig. 3b—the parameters are updated, and this random update lasts throughout the entire spiking period [21]. Given this inevitable variability in switching parameters, the consequent spiking dynamics undergoes variation, mainly alternating ISI in a spike train. Taking into account variation of switching parameters, following a normal distribution, ISI alternation in a single spike train could be predicted theoretically; interestingly, the results identified that the noise is featured by a Poisson-like noise—often seen in biological neurons [13]—and the uncorrelated spikes in the train [21].

The noise in the NLIF neuron inevitably elicits a noise in the gain function, which consequently leads to a noisy tuning function (e.g. Fig. 6a). Despite the presence of the noise, the average behaviour of the tuning function is comparable to the ideal one albeit noisy. This noisy tuning function brings about serious problems in information decoding. For instance, in Fig. 6b, the distributions of activities at three different stimuli (10, 20, and 30°) exhibit a remarkable overlap, so that discriminating two different stimuli from the neuronal activity is of difficult—especially if the observed activity is placed in the overlap. That is, a difficulty lies in decoding the neuronal response, implying information loss [1]. Several hypotheses on noise cancelling at the

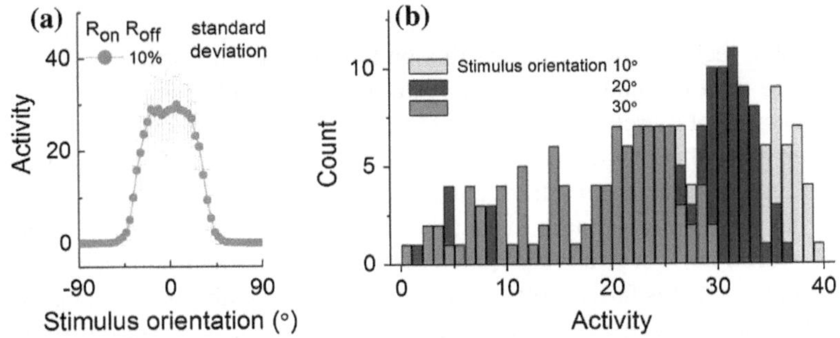

Fig. 6 **a** Tuning curve of the NLIF neuron with 10% R_{on} and R_{off} variability-tolerant threshold switches. The variability elicits large error bars compared with the ideal one in Fig. 5c. The distributions of activities at three different stimuli (10, 20, and 30°) represent large overlaps that lead neurons to difficulty in decoding the encoded information, i.e. orientation in this case

network scale have been proposed and theoretically identified their feasibility albeit not experimentally evidenced [47]. Probability of noise correlation and its positive effect on information decoding have also been proposed to reduce the uncertainty in stimulus discrimination to some extent [1, 2]. In addition, a recent study has theoretically proven that uncertainty in information decoding can be largely reduced by representation of a population of neurons [21]. All these hypotheses presume that information decoding can be conducted in a statistical manner; however, a biological mechanism for probabilistic coding has barely been understood for the time being.

As stated earlier, scalability of emerging artificial neurons is an important issue. The threshold switch is perhaps scaled down without remarkable degradation of its functionality [48]. Both R_{on} and R_{off} likely increase as the cell size shrinks so that materials engineering—able to adjust the resistance accordingly—needs to follow. However, making use of capacitors is a challenge. The capacitance determines the spiking dynamics in the way that it determines a time constant of the capacitance charging—immediately following spiking—and thus ISI. Provided the shrinkage of the capacitor area and the resulting decline in the capacitance, the spiking activity is seen at high-frequency scale and it may cause technical problems regarding the compatibility with the artificial synapse devices.

3 Summary and Outlook

Neuromorphic engineering provides versatile platform technologies that are possibly applied to various functional devices that recognize patterns in general such as visual, auditory, and semantic patterns. For the time being, a few neuromorphic products are already in the pipeline and expected to appear in the market in the near future [49, 50]. They are only a few examples in sight, however, a number of precious gems are probably hidden for the moment and waiting to be discovered. Gaining a better understanding of brain functionalities, e.g. recognition, perception, long-term memory, working memory, is an important premise for accelerating the progress in neuromorphic technologies. Towards this end, a significant portion of this chapter is devoted to the explanation as to biological neurons. Addressing artificial neuron models that are not merely delimited within the framework of hardware design is of importance given that the models are perhaps the roots of hardware artificial neurons.

Recalling the emphasis in Introduction, the target application determines the type of a neuron as well as a synapse in use; therefore, it is fruitful to diversify available artificial neuron models—ranging from a static and binary neuron to a dynamic and analogue neuron that is fairly biologically plausible. The latter may be most versatile in the light of its similarity to the biological neuron. This class of neurons follows the fidelity to the key features of the biological neuron—integrate-and-fire and variation in the response upon the input, i.e. gain function.

In designing a neuron, its detailed specifications need to be determined in combination with other significant components such as synapses and details of receiving data. Spiking neurons are required to allow synapses in connection to them to change their synaptic weights in the case that reweighting is supposed to occur. Thus, spiking dynamics, e.g. activity and spike height and width, should be compatible with the operational conditions of the synaptic weight change; otherwise, no weight change regardless of the activity of the adjacent neurons. For instance, use of an emerging memory element, e.g. phase-change memory and resistive switching memory, as an artificial synapse imposes remarkable constraints on the neuron design insomuch as the operational window of such an emerging memory is fairly narrow. In addition, when the target neuromorphic system deals with external dynamic stimulation—as our eyes do—the response rate should be comparable to the rate of the external stimulus. Otherwise, the cost of neuronal representation—approximately (the number of spikes) × (the energy consumption/spike)—outweighs the benefit, which is against one of the mottos of neuromorphic engineering: low power consumption. Thus, use of fairly large capacitors is inevitable regarding the reconciliation between low power consumption and high sensitivity to a time-varying stimulus. This serves as a significant obstacle to integrating the neuromorphic system, so that it is a challenge for the moment. A workaround solution may be to employ high dielectric constant materials representing much larger capacitance density at a given thickness compared to a conventional one such as SiO_x. Further technical issues in integration will appear with maturation of the technology, and multidisciplinary approaches to issues most likely provide shortcuts to the solutions.

Acknowledgements DSJ acknowledges the Korea Institute of Science and Technology grant (Grant No 2Z04510). DSJ also thanks Mr. Hyungkwang Lim for his kind help.

References

1. Averbeck, B.B., Latham, P.E., Pouget, A.: Nat. Rev. Neurosci. **7**, 358–366 (2006)
2. Averbeck, B.B., Lee, D.: J. Neurophysiol. **95**, 3633–3644 (2006)

3. Indiveri, G., Linares-Barranco, B., Hamilton, T.J., van Schaik, A., Etienne-Cummings, R., Delbruck, T., Liu, S.-C., Dudek, P., Häfliger, P., Renaud, S., Schemmel, J., Cauwenberghs, G., Arthur, J., Hynna, K., Folowosele, F., Saighi, S., Serrano-Gotarredona, T., Wijekoon, J., Wang, Y., Boahen, K.: Front. Neurosci. **5**, 1–23 (2011)
4. Hamann, C.H., Hamnett, A., Vielstich, W.: Electrochemistry, 2nd edn. WILEY-VCH Verlag GmbH & Co., Weinheim (2007)
5. Gadsby, D.C.: Nat. Rev. Mol. Cell Biol. **10**, 344–352 (2009)
6. Jeong, D.S., Kim, I., Ziegler, M., Kohlstedt, H.: RSC Adv. **3**, 3169–3183 (2013)
7. Malenka, R.C., Nicoll, R.A.: Science **285**, 1870–1874 (1999)
8. Dayan, P., Abbott, L.F.: Theoretical Neuroscience. The MIT Press, London (2001)
9. Butts, D.A., Goldman, M.S.: PLoS Biol. **4**, e92 (2006)
10. Deneve, S., Latham, P.E., Pouget, A.: Nat. Neurosci. **4**, 826–831 (2001)
11. Knill, D.C., Pouget, A.: Trends Neurosci. **27**, 712–719 (2004)
12. Ma, W.J., Beck, J.M., Latham, P.E., Pouget, A.: Nat. Neurosci. **9**, 1432–1438 (2006)
13. Calvin, W.H., Stevens, C.F.: Science **155**, 842–844 (1967)
14. Bialek, W., Rieke, F.: de Ruyter van Steveninck R. Warland D. Science **252**, 1854–1857 (1991)
15. Gerstner, W.: Neural Comput. **12**, 43–89 (2000)
16. Chance, F.S., Abbott, L.F., Reyes, A.D.: Neuron **35**, 773–782 (2002)
17. Eliasmith, C., Anderson, C.H.: Neural Engineering: Computation, Representation, and Dynamics in Neurobiological Systems. MIT Press, Cambridge, MA (2003)
18. Anderson, C.: Computational Intelligence Imitating Life. IEEE Press, New York (1994)
19. Pouget, A., Dayan, P., Zemel, R.S.: Annu. Rev. Neurosci. **26**, 381–410 (2003)
20. Weiss, Y., Fleet, D.J.: Velocity Likelihoods in Biological and Machine Vision. In Statistical Theories of the Cortex. MIT Press, Cambridge, MA (2002)
21. Lim, H., Kornijcuk, V., Seok, J.Y., Kim, S.K., Kim, I., Hwang, C.S., Jeong, D.S.: Sci. Rep. (2015)
22. Lapicque, L.: J. Physiol. Pathol. Gen. **9**, 620–635 (1907)
23. Abbott, L.F.: Brain Res. Bull. **50**, 303–304 (1999)
24. Gerstner, W., Kistler, W.M.: Spiking Neuron Models: Single Neurons, Populations. Cambridge University Press, Plasticity (2002)
25. Jolivet, R., Lewis, T.J., Gerstner, W.: Generalized Integrate-and-Fire Models of Neuronal Activity Approximate Spike Trains of a Detailed Model to a High Degree of Accuracy, vol 92. vol 2. doi:10.1152/jn.00190.2004 (2004)
26. Hodgkin, A.L., Huxley, A.F.: J. Physiol. **117**, 500–544 (1952)
27. FitzHugh, R.: Biophys. J. **1**, 445–466 (1961)
28. Nagumo, J., Arimoto, S., Yoshizawa, S.: Proc. IRE **50**, 2061–2070 (1962)
29. Destexhe, A.: Neural Comput. **9**, 503–514 (1997)
30. Kepler, T., Abbott, L.F., Marder, E.: Biol. Cybern. **66**, 381–387 (1992)
31. Mead, C.: Analog VLSI and Neural Systems. Adison-Wesley, Reading, MA (1989)
32. Mead, C.: Proc. IEEE **78**, 1629–1636 (1990)
33. Indiveri, G., Chicca, E., Douglas, R.: IEEE Trans. Neural Netw. **17**, 211–221 (2006)
34. Pickett, M.D., Medeiros-Ribeiro, G., Williams, R.S.: Nat. Mater. **12**, 114–117 (2012)
35. Crane, H.D.: Proc. IRE **50**, 2048–2060 (1962)
36. Jeong, D.S., Thomas, R., Katiyar, R.S., Scott, J.F., Kohlstedt, H., Petraru, A., Hwang, C.S.: Rep. Prog. Phys. **75**, 076502 (2012)
37. Jeong, D.S., Lim, H., Park, G.-H., Hwang, C.S., Lee, S.: Cheong B-k. J. Appl. Phys. **111**, 102807 (2012)
38. Ovshinsky, S.R.: Phys. Rev. Lett. **21**, 1450–1453 (1968)
39. Lee, M.-J., Lee, D., Cho, S.-H., Hur, J.-H., Lee, S.-M., Seo, D.H., Kim, D.-S., Yang, M.-S., Lee, S., Hwang, E., Uddin, M.R., Kim, H., Chung, U.I., Park, Y., Yoo, I.-K.: Nat. Commun. **4**, 2629 (2013)
40. Ahn, H.-W., Jeong, D.S., Cheong, B.-K., Kim, S.-D., Shin, S.-Y., Lim, H., Kim, D., Lee, S.: ECS Solid State Lett. **2**, N31–N33 (2013)
41. Mott, N.F.: Proc. Phys. Soc. Sect. A **62**, 416 (1949)

42. Cario, L., Vaju, C., Corraze, B., Guiot, V., Janod, E.: Adv. Mater. **22**, 5193–5197 (2010)
43. Han, J.-W.: Choi Y-K Bistable resistor (biristor)—gateless silicon nanowire memory. Symp. VLSI Technol. **15–17**(2010), 171–172 (2010). doi:10.1109/vlsit.2010.5556215
44. Pickett, M.D., Williams, R.S.: Nanotechnology **23**, 215202 (2012)
45. Liu, X., Sadaf, S.M., Son, M., Shin, J., Park, J., Lee, J., Park, S., Hwang, H.: Nanotechnology **22**, 475702 (2011)
46. Bienenstock, E., Cooper, L., Munro, P.: J. Neurosci. **2**, 32–48 (1982)
47. Deneve, S., Latham, P.E., Pouget, A.: Nat. Neurosci. **2**, 740–745 (1999)
48. DerChang, K., Tang, S., Karpov, I.V., Dodge, R., Klehn, B., Kalb, J.A., Strand, J., Diaz, A., Leung, N., Wu, J., Lee, S., Langtry, T., Kuo-wei, C., Papagianni, C., Jinwook, L., Hirst, J., Erra, S., Flores, E., Righos, N., Castro, H.: Spadini G A stackable cross point Phase Change Memory. In: IEEE International Electron Devices Meeting, vol. 7–9, pp. 1–4 (2009)
49. Merolla, P.A., Arthur, J.V., Alvarez-Icaza, R., Cassidy, A.S., Sawada, J., Akopyan, F., Jackson, B.L., Imam, N., Guo, C., Nakamura, Y., Brezzo, B., Vo, I., Esser, S.K., Appuswamy, R., Taba, B., Amir, A., Flickner, M.D., Risk, W.P., Manohar, R., Modha, D.S.: Science **345**, 668–673 (2014)
50. Gehlhaar, J.: Neuromorphic processing: a new frontier in scaling computer architecture. In: Architectural Support for Programming Languages and Operating Systems, Salt Lake City, UT, USA, pp. 317–318 (2014)

Synaptic Plasticity with Memristive Nanodevices

Selina La Barbera and Fabien Alibart

Abstract This chapter provides a comprehensive overview of current research on nanoscale memory devices suitable to implement some aspect of synaptic plasticity. Without being exhaustive on the different forms of plasticity that could be realized, we propose an overall classification and analysis of few of them, which can be the basis for going into the field of neuromorphic computing. More precisely, we present how nanoscale memory devices, implemented in a spike-based context, can be used for synaptic plasticity functions such as spike rate-dependent plasticity, spike timing-dependent plasticity, short-term plasticity, and long-term plasticity.

1 Introduction

There is nowadays an increasing interest in neuromorphic computing as a promising candidate to provide enhanced performances and new functionalities to efficient and low-power biomimetic hardware systems. On the one hand, seminal works in the 1950s with the concept of perceptron have been evolving continuously via software approaches. Starting from the simplest circuit structure (the perceptron), which corresponds to some formal neural networks representation, the Artificial Neural Networks (ANNs) have seen the emergence of very complex systems with impressive performances in recognition tasks, for example. Along these lines, the deep neural networks (DNNs) are today the most promising candidates for new computing systems. Even if the concepts of neurons and synapses are largely used in this field, a direct equivalence with their biological counterparts is not straightforward and sometimes impossible (or not biorealistic). On the other hand, recent progresses in neurosciences and biology have highlighted some basic mechanisms present in biological neural networks (BNNs). If the global understanding of the computing principle of such networks is out of reach, lots of key elements for computing have

S. La Barbera · F. Alibart (✉)
Institut d'Electronique, Microelectronique et Nanotechnologies,
UMR-CNRS 8520 Avenue H. Poincare, 59491 Villeneuve-d'Ascq, France
e-mail: fabien.alibart@iemn.univ-lille1.fr

© Springer (India) Pvt. Ltd. 2017

M. Suri (ed.), *Advances in Neuromorphic Hardware Exploiting Emerging Nanoscale Devices*, Cognitive Systems Monographs 31,
DOI 10.1007/978-81-322-3703-7_2

been evidenced. For example, spike timing-dependent plasticity (STDP), initially observed in BNNs, has attracted strong attention from computer science community since it opens the way to unsupervised learning systems which are expected to provide another breakthrough in the future of computing. In between these two main directions, i.e., ANNs and BNNs, neuromorphic computing and engineering emerge as an intermediate solution: The objective is still oriented toward the development of computing systems but with stronger analogy with biology with respect to ANNs. This classification should be carefully handled since the frontier between these different fields is far from being clear.

This chapter focuses on a crucial aspect addressed by neuromorphic computing: the synaptic plasticity. More precisely, starting from biological evidences, we will present some aspects of the synaptic plasticity that can be efficiently implemented with various emerging nanoscale memories for future biomimetic hardware systems.

2 Neuromorphic Systems: Basic Processing and Data Representation

By analogy with biological systems, information in neuromorphic systems is carried by spikes of voltage with a typical duration in the range of milliseconds. Starting from this simple observation, a first statement would be to consider neuromorphic networks as digital systems (spike being an all or nothing event). This direction was explored with the concept of neuron as logical unit performing logic operations in a digital way [32]. This short cut is of course hiding very important features observed in biological systems that present many analog properties of fundamental importance for computing. The first footprint of analog characteristics of biological systems can be simply emphasized by considering the analog nature of the synaptic connections bridging neurons. Analog synapses can be described in a first approximation as a tunable linear conductance, defining the synaptic weight between two neurons (this description is largely used in ANNs). Meanwhile, a more biorealistic description should consider the analog synapse as a complex device-transmitting signal in a nonlinear manner (i.e., frequency dependent). The second footprint of analog property is somehow embedded in the time-coding strategy used in BNNs: As the neuron is performing time integration of the digital spikes, the signal used for computing (the integrated value of the overall spiking activity) becomes an analog value regulating the spiking activity of the neuron. This second aspect is of particular relevance if we consider dynamical computing (i.e., natural data processing such as vision or sound that present a strong dynamical component). The temporal organization of spikes (or their time occurrence with respect to other spikes in the network) is carrying some analog component of the signal in biological networks. Now combining analog synapses with integrating neurons, the level of nonlinearity used by the network for computing the analog signal can be strongly modify. Simple linear filters can be realized with linear synaptic conductance associated with simple integrate-and-fire ($I\&F$) neurons or strongly nonlinear systems can be built, based on nonlinear synaptic conductance with complex integration at the neuron level such as leaky integrate-and-fire (*LIF*) or sigmoid neurons.

2.1 *Data Encoding in Neuromorphic Systems*

Starting from the statement that neuromorphic systems are analog systems, we have to define the appropriate data representation that will match the function to be realized. It should be stressed that data representation in biological systems is still under debate and a detail understanding is still a major challenge that should open new avenues from both a basic understanding and practical computing point of views.

Based on these general considerations, we can now try to present a simplified vision of data coding in biological systems that could be the basic ingredient for neuromorphic computing (i.e., hardware system implementation).

2.1.1 Rate-Coding Scheme

The simplest data representation corresponds to a rate-coding scheme, i.e., the analog value of the signal carrying information (or strength of stimuli) is associated with the average frequency of the train of pulse. The neuron can then transmit some analog signals through its mean firing rate. Rate-coding data representation is often used for static input stimuli representation but appears to be less popular for time-varying stimuli. Indeed, the sampling time interval $\Delta_{sampling}$ used for estimating the mean firing rate imply that events with fast temporal variation (typically variation on a timescale smaller than $\Delta_{sampling}$) cannot be described accurately. For example, the brain's time response to visual stimuli is around 100 ms and it cannot be accurately described in rate-coding systems that are typically in the range of frequencies from 1 to 100 Hz. A simple example of static data representation is to consider the representation of a static image from a $N \times M$ pixel array of black and white pixels into a $N \times M$ vector $X = (x_1, \ldots, x_i \ldots, x_n)$ where x_i can be either 0 or 1 (i.e., min and max frequencies). Then, this concept can be simply extended to analog data (such as pictures with different level of grays) by choosing properly the average firing rate.

2.1.2 Temporal-Coding Scheme

A second coding scheme is known as temporal coding in which each individual pulse of voltage is carrying a logical +1 and a time signature. This time stamp, associated with a given spike, can carry some analog value if we now consider its timing with respect to the other spikes emitted in the network [26]. The difficulty in this coding scheme is to precisely define the origin of time for a given spiking event that should depend on the event to be computed. A simple example is to consider a

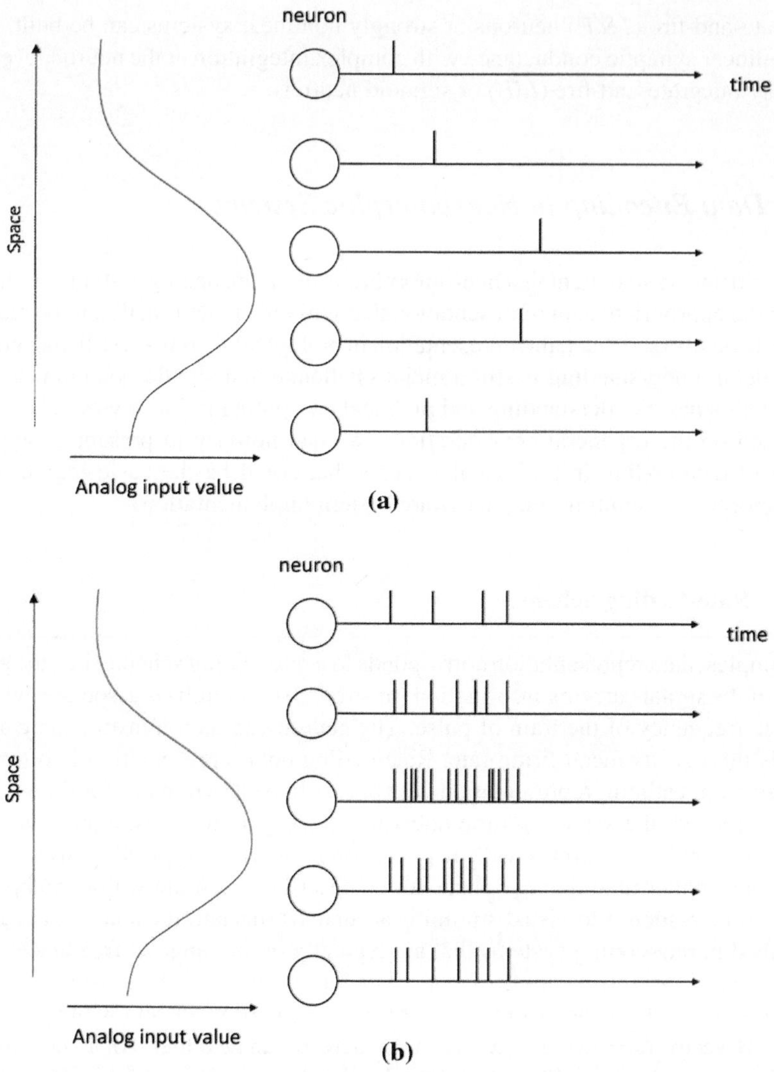

Fig. 1 Schematic illustration of data encoding schemes. A natural stimulus (such as a visual or auditory cue) is encoded through an input neuron population that sends and encodes the information on time in **a** time-coding scheme and in **b** rate-coding scheme

white point passing with a given speed in front of a detector with a black background and producing a pulse of voltage in each pixel of the detector when it is in front of it. By tracking both position of the activated pixel and time stamp attached to it, the dynamic of the event can be encoded.

Figure 1 shows how the rate- and time-coding schemes can be used to encode an analog signal x_i.

2.2 *Spike Computing for Neuromorphic Systems*

In this chapter, we will use only these two simplified data encoding concepts, but it should be stressed that other strategies such as stochastic-coding (i.e., the analog value of the signal is associated with the probability of a spike) are potential directions that deserve attention. We should also be aware that both rate coding and temporal coding have been evidenced to coexist in biological systems and both coding strategies can be used for powerful computing implementation. In fact, spike computing has attracted a large attention since the low-power performances of biological systems seem to be strongly linked to the spike coding used in such networks. But it should be emphasized and we should be aware of that translating conventional representation (i.e., digital sequences as in video) into spiking signal would most probably miss the roots of low-power computing in the biological system. Discretization of time and utilization of synchronous clock is in opposition with continuous time and asynchronous character of biological networks. Spike computing needs to be consider globally, i.e., by considering the full functional network and data encoding principle, from sensors to high-level computing elements. In this sense, recent development of bioinspired sensors such as artificial cochlea (sound detection) or artificial retinas (visual detection) with event-based representation opens many potentialities for fully spike-based computing where the dynamical aspect of spikes is naturally reproduced.

3 Synaptic Plasticity for Information Computing

By remaining in a computational spike-based context, we now focus on how a bioinspired network, composed in a first approximation of neurons and synapses, can process information (other functional units have to be considered if we want to describe precisely a biological networks such as proteins, glial cells, and ...). We can roughly categorized spike processing into (i) how spikes are transmitted between neurons, (ii) how spikes propagate along neurons, and (iii) how spikes are generated. These two last points can be attributed to 'neuron processing' and more precisely to the response of a biological membrane (the neuron membrane) to electrical or chemical signals. Many associated features such as signal integration, signal restoration, or spike generation are of first importance for spike computing, but these aspects are beyond the purposes of this chapter. The signal transmission will be the focus of this chapter, and different processes involved at the synaptic connection between two neurons will be described. We will concentrate on the dynamical responses observed in chemical synapses that are of interest for spike processing. Such synaptic mechanisms are broadly described as synaptic plasticity: the modification of the synaptic conductance as a function of the neurons activity. The specific synaptic weight values stored in the network are a key ingredient for neuromorphic computing. Such synaptic weight distribution is reached through synaptic learning and adaptation and can be described by the different plasticity rules present in the network. Furthermore, it should be noted that all the processes observed in biological synapses and their consequences on information processing are still an ongoing activity and final conclusions are still out of reach. Most probably, the efficiency of biological computing systems lies in a combination of many different features (restricted to the synapse level in this chapter) and our aim is to expose few of them that have been successfully implemented and to discuss their potential interest for computing.

In biology, synaptic plasticity can be attributed to various mechanisms involved in the transmission of the signal between pre- and post-synaptic neurons, such as neurotransmitter release modification, neurotransmitter recovery in the pre-synaptic connection, receptors sensitivity modification, or even structural modification of the synaptic connection (see [6]) for a description of the different mechanisms involved in synaptic plasticity).

It seems important at this stage to make a comprehensive distinction between different approaches used to describe the synaptic plasticity. The first approach, used to describe the synaptic plasticity, can be identified as a '*causal description*' based on the origin of the synaptic conductance modification. A second one is based on a '*phenomenological description*,' in which the temporal evolution (i.e., the dynamics) of the synaptic changes is the key element.

3.1 Causal Approach: Synaptic Learning Versus Synaptic Adaptation

By following the seminal idea of Hebb [19], a first form of plasticity is the so-called *synaptic learning* (Hebbian-type learning) and can be simply defined as an increase of the synaptic weight when the activity of its pre- and post-neuron increases. Many learning rules have been adapted following this simple idea of '*who fire together, wire together*.' Hebbian-type plasticity implies that the synaptic weight evolution dw_{ij}/dt depends on the product of the activity of the pre-neuron (a_i) and post-neuron (a_j) as follows:

$$\frac{dw_{ij}}{dt} \propto a_i \cdot a_j \tag{1}$$

This type of plasticity is defined in biology as *homosynaptic plasticity* [37]. Depending on the signal representation, i.e., rate coding or temporal coding, refinement (or particular cases) of Hebb's rule can be formulated such as spike rate-dependent plasticity (SRDP) or spike timing-dependent plasticity (STDP) with neuron activity defined as the mean firing rate or the spike timing, respectively.

A second form of synaptic plasticity can be referred to *Synaptic Adaptation* (where adaptation is in opposition with the notion of learning). In this case, synaptic weight modification depends on the activity of the pre- or post-neuron activity only or on the accumulation of both but in an additive process:

$$\frac{dw_{ij}}{dt} \propto a_i + a_j \tag{2}$$

In particular, if the synaptic plasticity depends only on post-activity, such mechanism is defined as *heterosynaptic plasticity* otherwise, if it is only pre-neuron activity dependent, it is named *transmitter-induced plasticity*.

Practically, this distinction seems very useful to classify the different synaptic processes that will be implemented and to evaluate their efficiency and contribution to the computing network performances. One major difficulty is that both *synaptic learning* and *synaptic adaptation* can manifest simultaneously and it becomes much more complicated in practical cases to make a clear distinction between them. In fact, learning in its large sense (i.e., how a network can become functional based on its past experiences) may involve both processes. Also, activity-independent weight modification can also be included to describe synaptic plasticity (e.g., to describe the slow conductance decay of inactive synapses, as it will be presented in the following paragraph).

3.2 Phenomenological Approach: Short-Term Plasticity Versus Long-Term Plasticity

Another important synaptic plasticity aspect that has to be considered is the *timescale* involved in the synaptic weight modification. Thus, by focusing on the synaptic plasticity dynamics observed in biological systems, synaptic weight modification can be either permanent (i.e., lasting for months to years) or temporary (i.e., relaxing to its initial state with a characteristic time constant in the milliseconds to hours range). This observation leads to the definition of *long-term plasticity* (LTP) and *short-term plasticity* (STP), respectively. We can notice that the boundary classification into long-term (LT) and short-term (ST) effects is not well defined and should be consider with respect to the task to be realized. Both STP and LTP can correspond to an increase or decrease of the synaptic efficiency, thus leading to the definition of facilitation (or potentiation) and depression, respectively. It is important to notice that there is no one to one equivalence between the concepts of STP, LTP, and the notion of short-term memory (STM) and long-term memory (LTM) which corresponds to a higher abstraction level (i.e., memory is then used in the sense of psychology). In this latter case, the information can be recalled from the network (i.e., information that has been memorized) and it cannot be directly associated with a specific set of synaptic weight with a given lifetime and plasticity rule. In fact, how synaptic plasticity can be related to the memorization of the information as well as how it is involved in different timescale of memory (from milliseconds to years) still remains debated.

4 Synaptic Plasticity Implementation in Neuromorphic Nanodevices

Many propositions of synaptic plasticity implementation with nanoscale memory devices have emerged these past years. By referring to the classification previously proposed, two main streams can be identified: the *causal* description and the *phenomenological* one. The first one relies on the implementation of the origin of the synaptic plasticity, without necessarily replicating the details of the spike transmission observed in biology. On the contrary, the second strategy has the aim to reproduce accurately the spike transmission properties observed in BNNs, by omitting the origin of the synaptic response, but rather by highlighting its temporal evolution.

In this section, we will present examples of practical devices implementation by following these two lines. Of course, a global approach based on a combination of both descriptions (the causal and the phenomenological one) would be the ideal solution to describe the synaptic weights distribution in ANNs for the future development of neuromorphic computing.

4.1 Causal Implementation of Synaptic Plasticity

In this first part, by following the *Causal* description, we will take into account the origin of the synaptic plasticity, without necessarily replicating the details of the spike transmission observed in biology.

4.1.1 Generality: Hebbian Learning

Hebbian learning has been at the basis of most of the learning strategies explored in neuromorphic computing. Hebbian-type algorithms define how a synaptic weight evolves during the learning experience and set the final weight distribution after the learning experience. Starting from its simplest form, i.e., '*who fire together, wire together*,' a first limitation of Hebbian learning can be evidenced. Indeed, if all synapses of the network are subject to Hebbian learning (Fig. 2), all synaptic connections should converge to their maximum conductivity after some time of activity since only potentiation is included in this rule, thus destroying the functionality of the network. A first addition to the Hebb's postulate is then to introduce anti-Hebbian plasticity that would allow to decrease the synaptic weight conductance (i.e., depression) when activity of both pre- and post-neurons are present (Fig. 2, green curve). One important consequence of this simple formulation (Hebbian and anti-Hebbian) is that the final synaptic weight distribution after learning should become bimodal (or binary), i.e., some weights became saturated to their maximum conductance (i.e., fully potentiated) while all the others should saturate to their lowest conductance state (i.e., fully depressed).

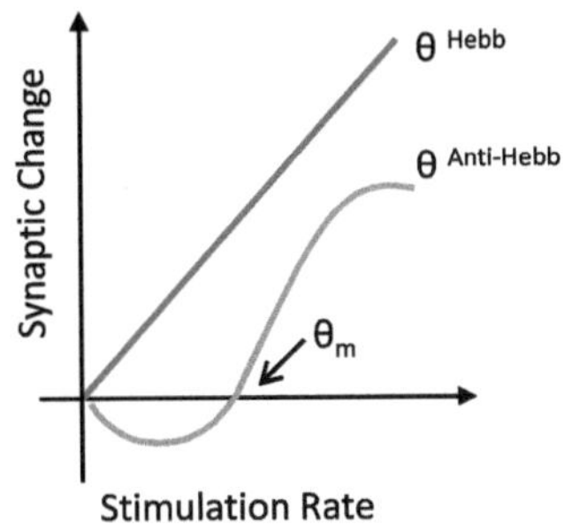

Fig. 2 Representation of the Hebbian rule (*purple*) and Hebbian/anti-Hebbian rule (*green*) for a constant post-neuron activity when pre-neuron activity is increased (stimulation rate). Addition of anti-Hebbian learning is a prerequisite in order to prevent all the synaptic weight to reach their maximal conductance

4.1.2 Time-Based Computing: Spike Timing-Dependent Plasticity

Without reviewing all the different STDP implementation in nanoscale memory devices propositions, we want to highlight some general ideas that are at the origin of this plasticity mechanism. The STDP was introduced by [2, 34] as a refinement of Hebb's rule. In this plasticity form (*Synaptic Learning*), the precise timing of pre- and post-synaptic spikes is taken into account as a key parameter for updating the synaptic weight. In particular, the pre-synaptic spike is required to shortly precede the post-synaptic one to induce potentiation, whereas the reverse timing of pre- and post-synaptic spike elicits depression. To understand how synaptic weights change according to this learning rule, we can focus on the process of synaptic transmission, depicted in Fig. 3.

Whenever a pre-synaptic spike arrives (t_{pre}) at an excitatory synapse, a certain quantity (r_1), for example, glutamate, is released into the synaptic cleft and binds to glutamate receptors. Such detector variable of pre-synaptic events r_1, increases whenever there is a pre-synaptic spike and decreases back to zero otherwise with a time constant τ^+. Formally, when $t = t_{pre}$ this gives the following:

$$\frac{dr_1}{dt} = -\frac{r_1(t)}{\tau_+} \tag{3}$$

We emphasize that r_1 is an abstract variable (i.e., state variable). Instead of glutamate binding, it could describe equally well some other quantity that increases after pre-synaptic spike arrival. If a post-synaptic spike arrives (t_{post}) at the same synapse, and the temporal difference with respect to the pre-synaptic one is not much larger than τ^+, the interaction between these two spikes will induce potentiation (LTP). As a consequence the synaptic weight $w(t)$ will be updated as follows:

$$w(t) = w(t) + r_1 \cdot A_2^+ \tag{4}$$

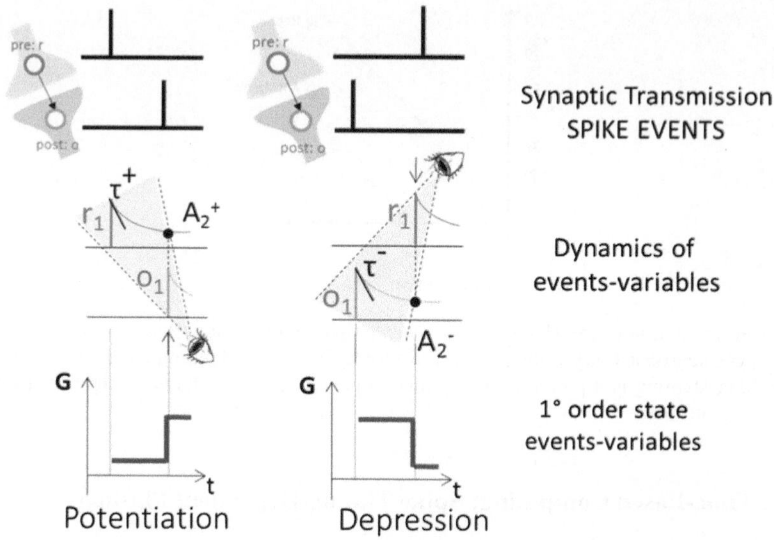

Fig. 3 Pair-based STDP learning rules: Long-term potentiation (LTP) is achieved thanks to a constructive pulses overlap respecting the causality principle (pre-before-post). On the contrary, if there is no causality correlation between pre- and post-synaptic spikes, long-term depression (LTD) is induced

If a pre-synaptic spike arrives after the post-synaptic one, another detector variable will be taken into account, relative to post-synaptic events (o_1), as shown in Fig. 3. Similarly, we can consider that the dynamics of o_1 can be described by time constant τ_{-}. Formally, when $t = t_{post}$ this gives the following:

$$\frac{do_1}{dt} = -\frac{o_1(t)}{\tau_{-}} \tag{5}$$

If the temporal difference is not much larger than τ^{-}, the spike interaction will induce depression (LTD). As a consequence the synaptic weight $w(t)$ will be updated as follows:

$$w(t) = w(t) - o_1 \cdot A_2^{-} \tag{6}$$

One of the important aspects of STDP is to present both Hebbian and anti-Hebbian learning. Replicating the exact biological STDP window (Fig. 4a) is not a mandatory condition for implementing interesting learning strategies (other shapes have been reported in biology) while balancing the Hebbian/anti-Hebbian contribution remains a challenge in order to maintain STDP learning stable. It should be noted that synaptic weight distribution becomes bimodal after some time of network activity if this simple STDP window is implemented [40].

The proposition of memristor [38] provides an interesting framework for the implementation of synaptic weights (i.e., analog property of the memory) and for the

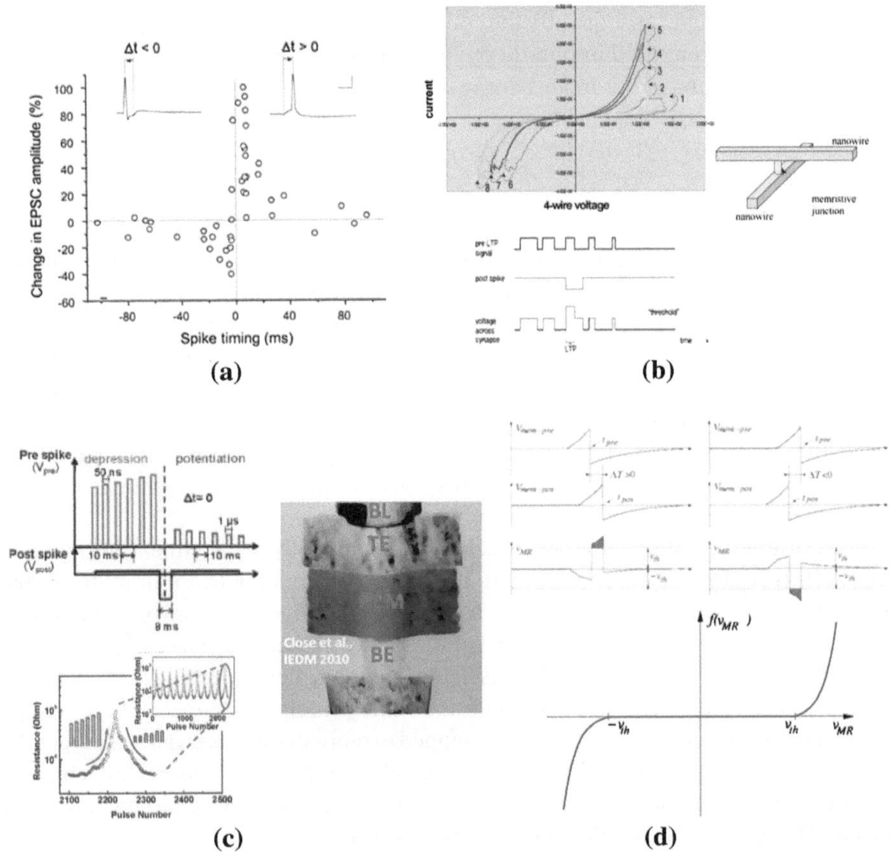

Fig. 4 **a** Biological STDP window from [4]. In all three cases: **b–d**, the particular shape of the signal applied at the input (pre-neuron) and output (post-neuron) of the memory element induces a particular effective voltage that induces potentiation (increase of conductance) or depression (decrease of conductance) reproducing the STDP window of (**a**). **b** First proposition of STDP implementation in nanoscale bipolar memory devices where time multiplexing approach was considered. In this case, the STDP window can be reproduced with high fidelity while the spike signal is far from biorealistic. **c** Implementation of STDP in unipolar PCM devices. Still the STDP window can be reproduced precisely while the signal is not biorealistic. **d** Proposition of STDP implementation with bipolar memristor. Both the STDP window and pulse shape are mapped to biorealistic observations

implementation of STDP in particular. Nanoscale memories or 'memristive devices,' as previously introduced, are electrical resistance switches that can retain a state of internal resistance based on the history of applied voltage and the associated memristive formalism. Using such nanoscale devices provides a straightforward implementation of this bioinspired learning rule. In particular, the modulation of the memristive weight (i.e., the conductance change $\Delta G(W, V)$ is controlled by an internal parameter W that depends on the physics involved in the memory effect. In most of the memory technologies used for such bioinspired computational purpose, the internal state variable W (and consequently the conductance) is controlled through the applied voltage or the current (and implicitly by its duration). Mathematically, this behavior corresponds to a first-order memristor model:

$$\frac{dW}{dt} = f(W, V, t) \tag{7}$$

with $I = V \cdot G(W, V)$. Practically, by exploiting memristive devices as synapses, most of the STDP implementation relies on specific engineering of the spikes's shape that convert the time correlation (or anti-correlation) between pre- and post-spikes into a particular voltage that induces a modification of the memory element conductance. The time lag induced by pre-synaptic events, as the r_1 variable in Fig. 3, determines that the potentiation is converted into a particular voltage across the memristor in order to induce an increase of conductance when a post-synaptic spike interact with it. Similarly, time lag induced by post-synaptic events in analogy with o_1 variable in Fig. 3 will induce depression in form voltage across the memristor when interacting with a pre-synaptic spike.

First implementation was proposed by Snider [36] with time multiplexing approach (Fig. 4b), in which, although the spike signal is far from biorealistic, the STDP window can be reproduced with high fidelity. Figure 4c shows another successful STDP implementation with non-biorealistic signal in a phase-change memory device [22]. Depending on the particular memory device considered, different encoding strategies were proposed with the same principle of input/output voltage correlation in which the STDP window mapped to biorealistic observations. Recently, by going deeper in the memristive switching behavior (i.e., by considering a higher-order memristive model), STDP was proposed through even more biorealistic pulse shape [21], as it will be explained in the Sect. 4.1.4.

4.1.3 Rate-Based Computing: The BCM Learning Rule

While the STDP learning rule has been largely investigated these past years, another refinement of the Hebb's rule can be formulated in the case of rate-coding approaches. Bienenstock et al. [5] proposed in the 1980s the BCM learning rule with the concept of '*sliding threshold*' that ensures to maintain the weight distribution bounded and thus avoiding unlimited depression and potentiation resulting from simple Hebbian learning implementation. The BCM learning rule can be simply formalized as follows:

$$\frac{dw_{ij}}{dt} = \varphi(a_j(t)) \cdot a_i(t) - \varepsilon w_{ij} \tag{8}$$

where w_{ij} is the synaptic conductance of the synapse bridging the pre-neuron i and post-neuron, j, a_i, and a_j are the pre- and post-neuron activities, respectively, ε is a constant related to a slow decaying component of all the synaptic weights (this term

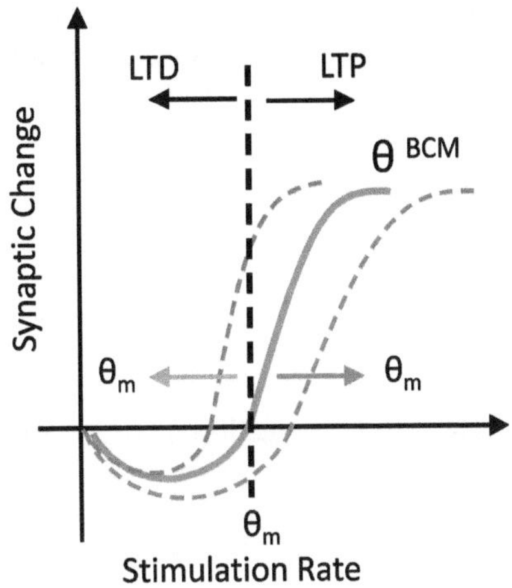

Fig. 5 BCM learning rule representation. The synaptic weight modification is represented as a function of pre-neuron activity for a fixed post-neuron activity. The sliding threshold depends on the mean post-neuron activity, i.e., θ_m is increased if a_j increases while θ_m is decreased if a_j decreases, thus preventing unlimited synaptic weight modification

appears to become important in special cases, see [5] but not mandatory) and φ a scalar function parametrized as follows:

$$\varphi(a_j) < 0 \; for \; a_j < \theta_m \quad \& \quad \varphi(a_j) > 0 \; for \; a_j > \theta_m$$

where θ_m is a threshold function that depends on the mean activity of the post-neuron. A first-order analysis can be realized on this simple learning rule. (i) Both Hebbian-type learning (product between a_i and a_j) and adaptation (through the small decay function that is not related to pre- and post-neuron activities) are present in this rule. (ii) The threshold ensures that both Hebbian and anti-Hebbian plasticity can be obtained through the scalar function φ that can take positive and negative values (potentiation and depression). (iii) Thus, the '*sliding threshold effect*' corresponds to the displacement of the threshold as a function of the post-neuron activity and is a key ingredient to prevent the synaptic weight distribution to become bimodal. Indeed, if the mean post-neuron activity is high, any pre-neuron activity should induce potentiation (most probably). If now θ_m is increased when the mean post-neuron activity increases, it will increase the probability of depression or at least reduce the magnitude of potentiation and consequently limit the potentiation of the weight (Fig. 5).

The BCM learning rule was initially proposed for rate-coding approaches and was measured in BNNs in the long-term regime of the synaptic plasticity. The BCM learning rule has been shown to maximize the selectivity of the post-neuron [5]. Only few works have demonstrated partially the BCM rule in nanoscale memory devices with some limitations. Lim et al. [25] proposed to describe the weight saturation in TiO_2 electrochemical cells subject to rate-based input. This work demonstrated the sliding threshold effect describing the saturation of the weight during potentiation and depression but did not reproduce the Hebbian/anti-Hebbian transition. Ziegler et al. [47] demonstrate the sliding threshold effect in the long-term regime but without considering explicitly a rate-coding approach, i.e., neuron activity was simply associated with the pre- and post-neuron voltages. Kim et al. [21] proposed an adaptation of the BCM rule in second-order memristor, as it will be presented in the next section, but in a transmitter-induced plasticity context, thus missing the Hebbian-type plasticity initially proposed in the BCM framework. Future works are expected to provide stronger analogy with BCM rule, both from a phenomenological point of view (i.e., biorealistic rate-coding implementation) and from a causal point of view (i.e., reproducing all the aspects of the BCM rule).

4.1.4 Reconciliation of BCM with STDP

On the one hand, the importance of individual spikes and their respective timing can only be described in the context of STDP. The time response in the visual cortex being in the order of 100 ms, rate-coding approaches are unlikely to offer a convenient description of such processes while time coding could. On the other hand, simple STDP function misses the rate-coding property observed in BNNs and conveniently described in the context of the BCM. More precisely, in the case of pair-based STDP, both potentiation and depression are expected to decrease as the activity mean frequency of the network is increased while BNNs show opposite trend. Izhikevich et al. [20] proposed that classical pair-based STDP, implemented with the nearest-neighbor spike interactions, can be mapped to the BCM rule. However, their model failed to capture the frequency dependence [35] if pairs of spikes are presented at different frequencies [14].

From a neurocomputational point of view, Gjorgjieva et al. [18] proposed a triplet STDP model based on the interactions of three consecutive spikes as generalization of the BCM theory. This model is able to describe plasticity experiments that the classical pair-based STDP rule has failed to capture and is sensitive to higher-order spatio-temporal correlations, which exist in natural stimuli and have been measured in the brain. As done for the pair-based case, to understand how synaptic weights change according to this learning rule, we can focus on the process of synaptic transmission, depicted in Fig. 6.

Instead of having only one process triggered by a pre-synaptic spike, it is possible to consider several different quantities, which increase in the presence of a pre-synaptic spike. We can thus consider, r_1 and r_2 two different detectors variables of pre-synaptic events and their dynamics can be described with two time constant τ_+ and τ_x ($\tau_x > \tau_+$). Formally, when $t = t_{pre}$, this gives the following:

$$\frac{dr_1}{dt} = -\frac{r_1(t)}{\tau_+} \quad \& \quad \frac{dr_2}{dt} = -\frac{r_2(t)}{\tau_x} \tag{9}$$

Similarly, we can consider, o_1 and o_2 two different detector variables of post-synaptic events and their dynamics can be described with two time constants τ_- and τ_y ($\tau_y > \tau_-$). Formally, when $t = t_{post}$, this gives the following:

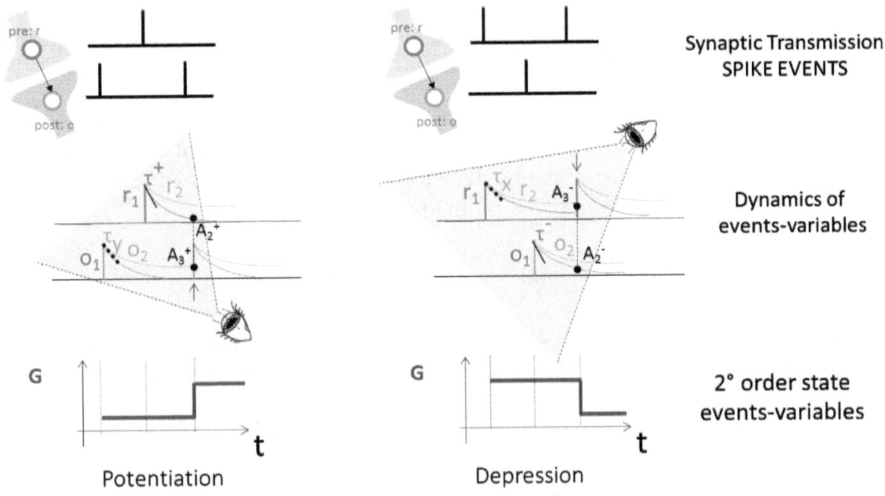

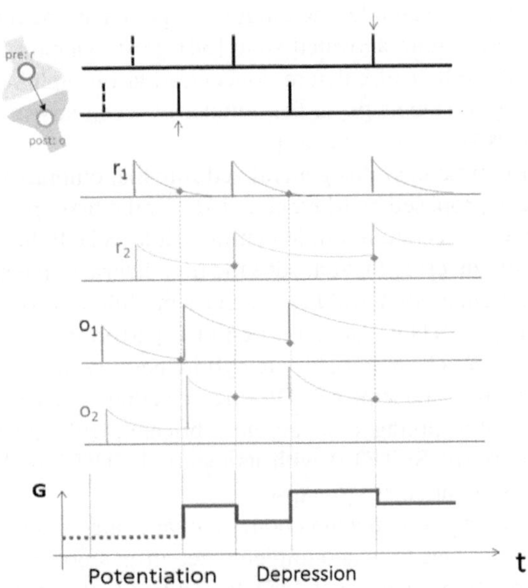

Fig. 6 Triplet-based STDP learning rules. **a** Synaptic weight potentiation (LTP) is achieved thanks to (post-pre-post) spike iterations, as a result the relative time lag of the detector-variable dynamics. Similarly a synaptic weight depression (LTD) is induced with (pre-post-pre) spike interactions. **b** Synaptic weight evolution in function of time correlation of pre- and post- spikes

$$\frac{do_1}{dt} = -\frac{o_1(t)}{\tau_-} \quad \& \quad \frac{do_2}{dt} = -\frac{o_2(t)}{\tau_y} \tag{10}$$

We assume that the weight increases after post-synaptic spike arrival by an amount that is proportional to the value of the pre-synaptic variable r_1 but depends also on the value of the second post-synaptic detector o_2. Hence, post-synaptic spike arrival at time t_{post} triggers a change given by the following:

$$w(t) = w(t) + r_1(t) \cdot (A_2^+ + A_3^+ o_2(t)) \tag{11}$$

Similarly, a pre-synaptic spike at time t_{pre} triggers a change that depends on the post-synaptic variable o_1 and the second pre-synaptic variable r_2 as follows:

$$w(t) = w(t) - o_1(t) \cdot (A_2^- + A_3^- r_2(t)) \tag{12}$$

As done previously, we emphasize that r_1, r_2, o_1, and o_2 are abstract variables that not identify with specific biophysical quantities. Biological candidates of detectors of pre-synaptic events are, for example, the amount of glutamate bound [9] or the number of NMDA receptors in an activated state [34]. Post-synaptic detectors o_1 and o_2 could represent the influx of calcium concentration through voltage-gated Ca^{2+} channels and NMDA channels [9] or the number of secondary messengers in a deactivated state of the NMDA receptor [34].

A possible solution to implement this generalized rule that embraces both BCM theory and STDP has been proposed by Mayr et al. [31] for the first time in $BiFeO_3$ memristive devices. They succeeded in implementing triplet STDP through a more complex spikes's shape engineering that encodes the time interaction between more than two pulses into a particular voltage able to induce a modification of the memory element conductance. Triplet STDP rule has been also performed by Williamson et al. [43] in asymmetric TiO_2 memristor in hybrid neuron/memristor system. Subramaniam et al. [39] have used triplet STDP rule in a compact electronic circuit in which neuron consists of a spiking soma circuit fabricated with nanocrystalline-silicon thin-film transistors (ns-Si TFTs) with nanoparticle TFT-based short-term memory device and HfO_2 memristor as synapse.

Another generalized description, in which both time- and rate-coding approaches are taken into account at the same time and implemented in an amorphous InGaZnO memristor, has been proposed by Wang et al. [42]. In addition to the conventional ion migration induced by the application of pulse of voltage, another physical mechanism of the device operation occurs: the gradient of the ions concentration, leading to the appearance of ion diffusion, resulting in an additional state variable. Kim et al. [21] recently proposed a second-order memristor that offers an interesting solution toward this goal of reconciliation of various learning mechanisms in a single memory device.

Mathematically, in analogy to the previous definition, a second-order memristor model can be described as follows:

$$\frac{dW_1}{dt} = f_1(W_1, W_2, V, t) \quad \& \quad \frac{dW_2}{dt} = f_2(W_1, W_2, V, t) \tag{13}$$

with $I = V \cdot G(W_1, W_2, V, t)$ and implemented with a simple nonoverlapping pulses protocol for the synaptic weight modulation.

The interest behind this higher-order memristor description is to provide additional parameters that will ensure some other higher-order interaction between pulses (i.e., more than two), while the pair-based interaction is preserved. More precisely, as shown in Fig. 7a, the temperature has been proposed as second-order state variable that exhibits short-term dynamics and naturally encodes information on this relative timing of synapse activity. By exploiting these two state variables (i.e., the conductance and the temperature), STDP has been implemented, as it is shown in Fig. 7a. Specifically, the first 'heating' spike elicits an increase in the device temperature by Joule effect regardless of the pulses polarity, which then tends naturally to relax after the removal of the stimulation, then temporal summation of the thermal effect can occur and can induce an additional increment in the temperature of the device if the second 'programming' spike is applied before T has decayed to its resting value.

Longer time interval will induce a small conductance change because of the heat dissipation responsible to a lower residual T when the second spike is applied. Thus, the amount of the conductance change (long-term dynamics) can be tuned by the relative timing of the pulses encoded in the short-term dynamics of second *state variable* (i.e., the temperature T).

Du et al. [17] have proposed another second-order memristor model. Also in this case, two state variables are used to describe an oxide-based memristor. The first one, as in the previous example, directly determines the device conductance (i.e., the synaptic weight). Specifically, this first state variable represents the area of the conducting channel region in the oxide memristor, thus directly affecting the device conductance. The second state variable represents the oxygen vacancy mobility in the film which directly affects the dynamics of the first state variable (conductance) but only indirectly modulates the device conductance (Fig. 7a). Equivalently to T, the w is increased by application of a pulse and then tends to relax to an initial value and affects the first state variable by increasing the amount of conductance change in a short timescale. By exploiting this second-order memristor model, Du et al. [17] have demonstrated that STDP can be implemented in oxide-based memristor by simple nonoverlapping pre- and post-synaptic spike pairs, rather than through the engineering of the pulse's shape (Fig. 7b).

In neurobiology, the timing information is intrinsically embedded in the internal synaptic mechanisms. Malenka and Bear [27] have demonstrated that together with the neurotransmitter dynamics in the presynaptic connection, secondary internal state variables, such as the natural decay of the post-synaptic calcium ion (Ca^{2+}) concentration, are involved in the synaptic weight modulation and the synaptic plasticity that can be achieved by simple nonoverlapping spikes and tuned by synaptic activity (i.e., rate- and timing-dependent spikes) which brings an interesting analogy between biological processes and material implementation described above [18].

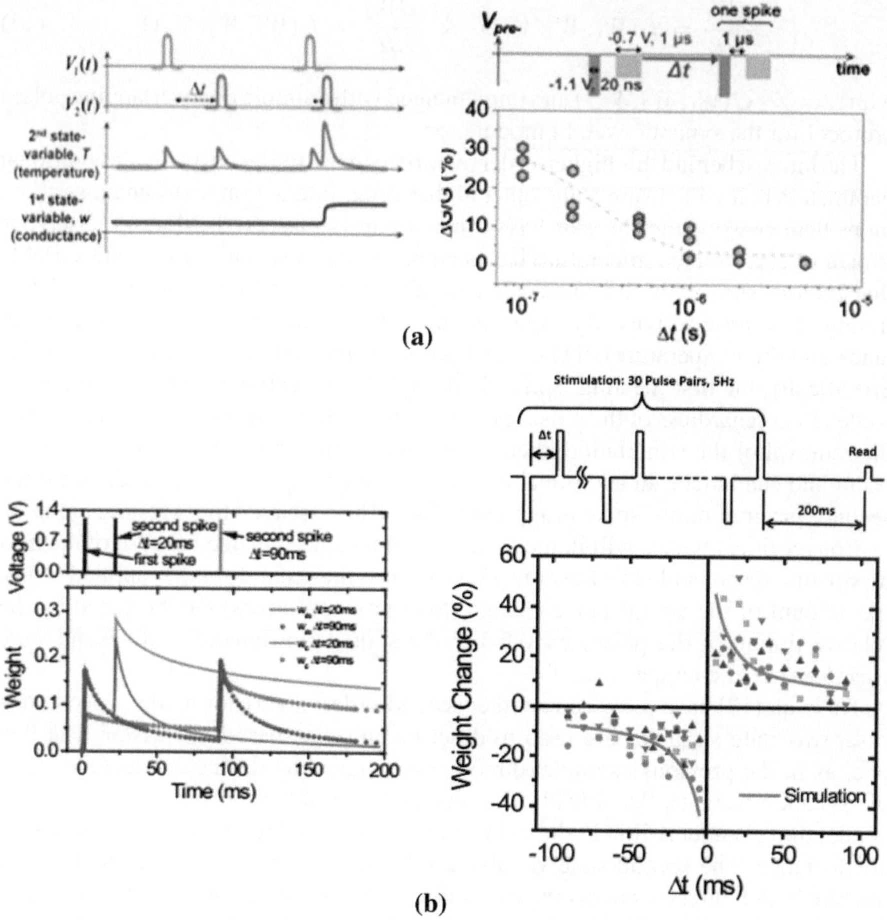

Fig. 7 Second-order memristor model. **a** On the right: the modulated second-order state variable exhibits short-term dynamics and naturally encodes information on the relative timing and synapse activity. On the *left* STDP implementation: memristor conductance change as a function of only two spikes (i.e., each spike consists of a programming pulse and a heating pulse) [21]. **b** On the *right* Simulation results illustrating how the short-term behavior affected long-term weight change. The difference in long-term weight is caused by the different values of residue of the second state variable at the moment when the second pulse is applied. The first and the second state variables under two conditions (interval between two pulses $\Delta t = 20, 90\,\text{ms}$) are shown. On the *left* memristor weight change as a function of the relative timing between the pre- and post-synaptic pulses without pulses overlapping (STDP implementation) [17]

The hypothesis that several synaptic functions manifest simultaneously and are interrelated at synaptic level seems accepted by different scientific communities. Recent biological studies indicate that multiple plasticity mechanisms contribute to cerebellum-dependent learning [8]. Multiple plasticity mechanisms may provide the flexibility required to store memories over different timescales encoding the dynamics involved. From a computational point of view, Zenke et al. [46] have recently proposed the idea to use multiple plasticity mechanisms at different timescales. Instead of focusing on particular and local learning schemes, their strategy aims to create memory and learning functions through interplay of multiple plasticity mechanisms. By following this trend of multi-scale plasticity mechanisms, Mayr et al. [30] have realized a VLSI implementation in which short-term, long-term, and meta-plasticity interact each other at different timescales to tune the overall synapse weights distribution.

4.2 Phenomenological Implementation of Synaptic Plasticity

In this section, we will follow the second synaptic description approach: the *phenomenological* one. The spike transmission properties observed in BNNs will be presented as a function of the temporal evolution of the synaptic weight.

4.2.1 STP in a Single Memristive Nanodevices

As previously mentioned, the *transmitter-induced plasticity* is a particular form of synaptic adaptation that depends only on pre-neuron activity. From a phenomenological point of view, such plasticity is most often observed on short timescale, thus belonging to the class of STP. As shown in Fig. 8b, this STP regime is frequency dependent and can be used to modulate the synaptic weights distribution as a function of network activity. From a biological view point, a phenomenological model of frequency-dependent synaptic transmission was used to describe such synaptic response in STP regime [28]. The primary synaptic parameters are the absolute synaptic efficacy (A), the utilization of synaptic efficacy (U), recovery from depression (τ_{rec}), and the recovery from facilitation (τ_{facil}) (Fig. 8a). In this model, synaptic response is then dependent on the finite amount of neurotransmitter resources in the pre-synaptic neuron and their respective dynamics (utilization and recovery) and on the absolute efficacy of the synaptic connection which could depends on post-synaptic neuron receptors sensitivity or synaptic connection, for example. The most likely biophysical mechanisms underlying changes in the value of these synaptic parameters were considered [28].

If we consider a temporal-coding approach in which pulses are considered as discrete events, STP can be evidenced through the notion of paired pulse facilitation (PPF) corresponding to the enhancement of a pulse transmission when this latter closely follows a prior impulse. The countereffect (i.e., corresponding to depression) is referred to as paired pulse depression (PPD). If we now focus on rate-coding approaches, facilitation and depression can be simply described as a high-pass and low-pass filters. Depending on the mean firing rate of the synapse, signal can be enhanced or depressed when pre-neuron frequency is increased. A simple material implementation of such mechanism can be realized through passive RC circuits.

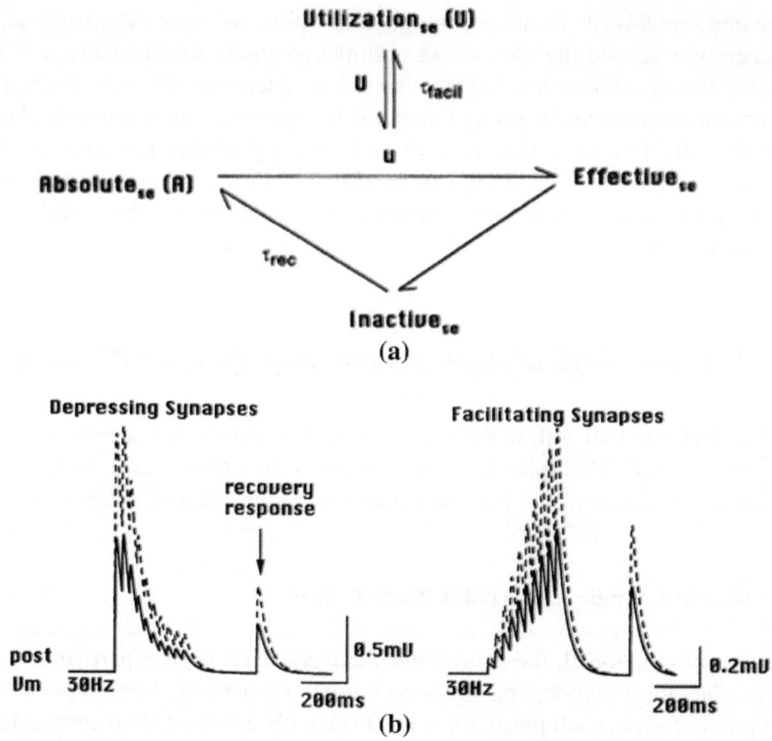

Fig. 8 Phenomenological model of frequency-dependent synaptic transmission. **a** Each AP utilizes U a fraction of the available/recovered synaptic efficacy R. When an AP arrives, U is increased by an amplitude of u and becomes a variable, $U1$. **b** Phenomenology of changing absolute synaptic efficacy parameter A. On the *left* synaptic responses of depressing synapses when A is increased 1.7-fold. On the *right* synaptic responses of facilitating synapses when A is increased 1.7-fold. Adapted from [29]

It turns out that RC circuits with time constants in the milliseconds to seconds range leads to very high capacity with large area (even at low current operation) that are a severe limitation for hardware implementation of STP. Different alternative approaches can realize more efficiently such dynamical effects by taking advantage of physical mechanisms present in nanoscale memory devices.

The first proposition of STP with nanodevices was realized in a nanoparticles/organic memory transistor (NOMFET) [3]. The basic principle of this device is equivalent to a floating gate transistor. Charges, stored in the nanoparticles, modify the channel conductivity via coulomb repulsion between the carriers (holes) and the charged nanoparticles. The particularity of this device relies on the leaky memory behavior: Charges stored in the nanoparticles tend to relax with a characteristic time constant in the 100–200 ms range [16]. When the NOMFET is connected in a diode-like configuration (Fig. 7a), each input spike (with a negative voltage value) charges the nanoparticles and decreases the NOMFET conductivity. Between pulses,

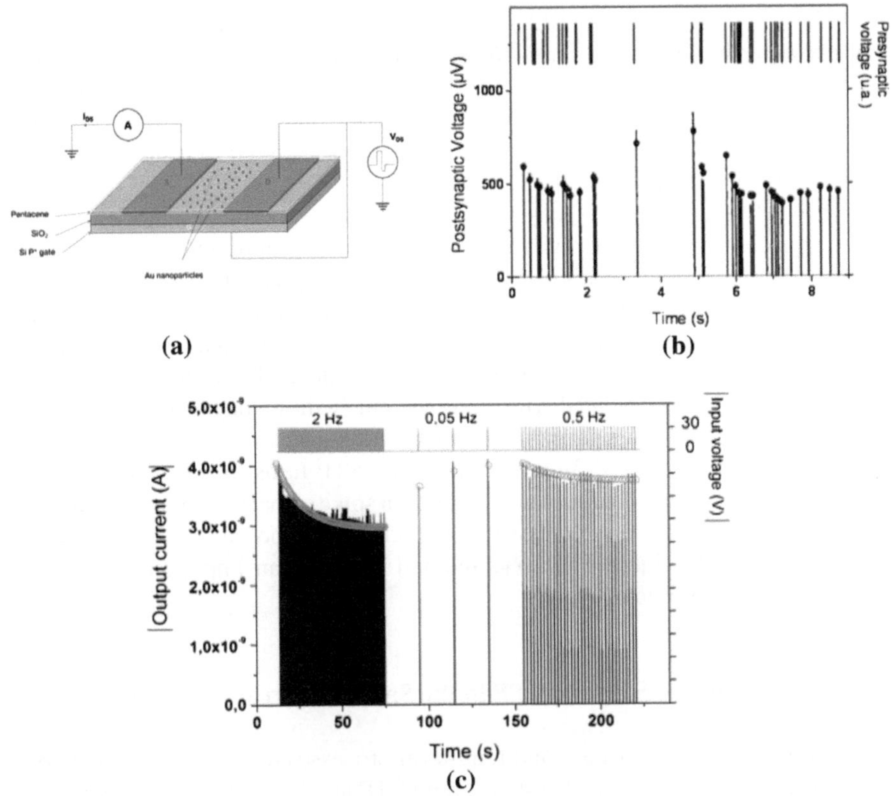

Fig. 9 STP implementation in a NOMFET. **a** Schematic representation of the NOMFET and pseudo-two-terminal connections of the device. **b** Comparison between the frequency-dependent post-synaptic potential response of a depressing synapse (*lines*) and the iterative model of Varela et al. (*dots*), adapted from [41], as a function of frequency of the pre-synaptic input signal. **c** Response (drain current) of NOMFET with L/W ratio of 12 μm/113 μm and NP size of 5 nm to sequences of spikes at different frequencies (pulse voltage $Vp = -30\,\mathrm{V}$)

charges escape from the nanoparticles and the conductivity relaxes toward its resting value. By analogy with biology, this device mimics the STP observed in depressing synapses (Fig. 9) and described by [1]. As a matter of comparison, this synaptic functionality is realized with a single memory transistor while its implementation in Si-based technologies (i.e., CMOS) required 7 transistors [7].

STP has been also demonstrated in two-terminal devices that would ensure higher devices density when integrated into complex systems. Equivalently, STP in two-terminal devices is implemented by taking advantage of the volatility of the different memory technologies (i.e., low retention of the state that is often a drawback in pure memory applications). Redox systems based on electrochemical memory cell (ECM) [33] or valence change memory (VCM) [12, 44] have demonstrated STP with a facilitating behavior. In such devices, short-term plasticity is ensured by the low stability of the conducting filaments that tends to dissolve, thus relaxing the device toward the insulating state. TiO_2 VCM cells have been reported with both facilitating and depressing behavior [25] with relaxation related to oxydo-reduction counter-reaction. Protonic devices have demonstrated STP with depressing functionality due to proton recovery latency from atmosphere required to restore the proton concentration and conductivity [15].

In terms of functionality, [1] has demonstrated that depressing synapses with STP act as a gain control device (at high frequency, i.e., high synaptic activity, the synaptic weight is decreased, thus leading to a lowering of the signal when activity becomes too important). More generally, STP (both depressing and facilitating) provides a very important frequency coding property (as depicted in Fig. 7 that could play a major role in the processing of spike rate-coded information). Indeed, if a simple integrate-and-fire neuron ($I\&F$) is associated with static weight (with no dependence with spike frequency), the computing node (i.e., neuron and synapses) is only a linear filter (linear combination of the different input) while STP turns the node to nonlinear. This property (i.e., locally induced nonlinearity in spike signal transmission) has been used to implement reservoir-computing approaches as proposed by Buonomano and Maass [10] with the liquid-state machine and could be an important property of biological systems for computing.

4.2.2 Coexistence of STP and LTP in the Same Memristive Nanodevice

If the contribution of short-term and long-term processes to computing is not completely understood in biological systems, both STP and LTP effects in synaptic connections have been evidenced and should play a crucial role. A first approach is to consider that repetition of short-term effects should lead to long-term modification in the synaptic connections. This behavior would explain the important hypothesis of memory consolidation in the sense of psychology [24]. Ohno et al. [33] reported for the first time the transition from short-term to long-term potentiation in atomic bridge technology (Fig. 10). Considering again the *transmitter-induced plasticity* dependent on the pre-synaptic activity (associated with spike rate in this case), the synaptic conductivity is increased due to the formation of a silver (Ag) filament across the insulating gap. While for low frequency, the bridge tends to relax between pulses; higher frequencies lead to a strong filament that maintains the device in the ON state. These results suggest a critical size of the bridging filament in order to maintain the conductive state stable (i.e., providing a LTP of the synaptic connection).

Similar results have been obtained in a variety of memory devices where filamentary switching displayed two regimes of volatility. Wang et al. [42] have shown that STP-to-LTP transition can occur through repeated 'stimulation' training. By stimulating sequentially an oxide-based memristive device with 100 positive pulses, the synaptic weight gradually increases with the number of pulses. Once the applied voltage is removed, a spontaneous decay of synaptic weight occurs in the case of no

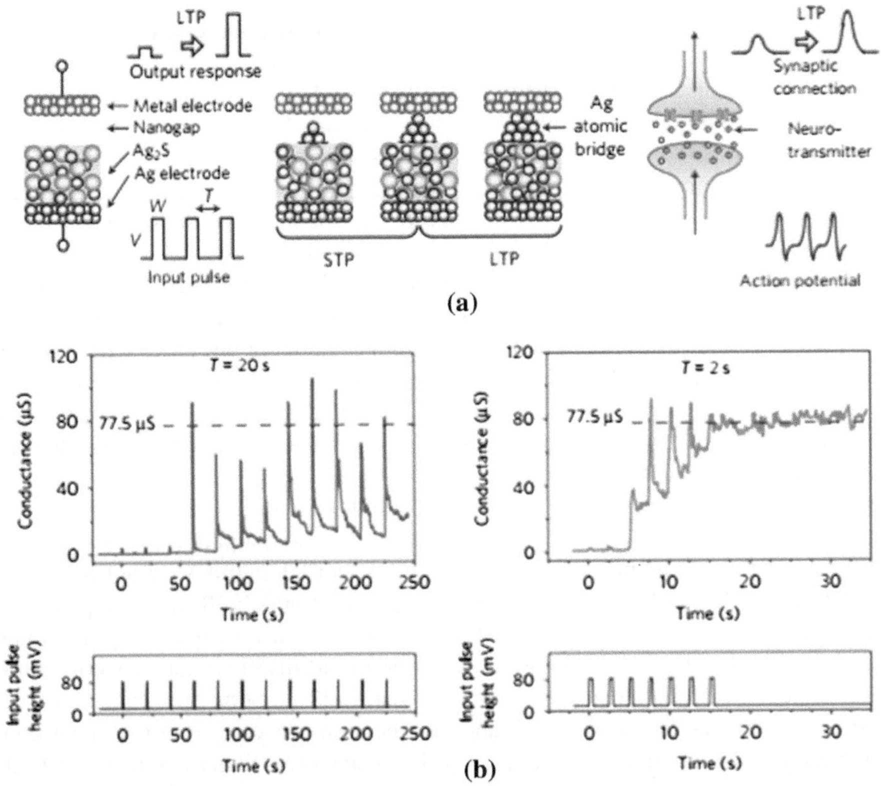

Fig. 10 STP and LTP implementation in an ECM cell depending on input pulse repetition time. **a** Schematic representation of the Ag_2S ECM cell and the signal transmission of a biological signal. Application of input pulses causes the precipitation of Ag atoms from the Ag_2S electrode, resulting in the formation of an Ag atomic bridge between the Ag_2S electrode and a countermetal electrode. When the precipitated Ag atoms do not form a bridge, the ECM cell works in the STP regime. After an atomic bridge is formed, it works as LTP. **b** Frequent stimulation ($T = 2$ s) causes long-term enhancement in the strength of the synaptic connection while short-term enhancement is induced at lower frequency ($T = 20$ s) [33]

external inputs. The synaptic weight does not relax to the initial state, but stabilizes at a mid-state, which means that the change of synaptic weight consists of two parts: STP and LTP.

Chang et al. [13] have evidenced a continuous evolution of the volatility as a function of the conductivity level of the device in WO_3 oxide cells attributed to the competition between oxygen vacancies drift (creation of conductive path across the device) and lateral diffusion (disruption of the conducting filament). Another description of these two regimes of volatility could be associated with a competition between surface and volume energies in the conductive filament [45].

4.2.3 Conflict Between Causal and Phenomenological Description

If this concept of ST-to-LT transition has been well demonstrated in a variety of nanoscale memory devices, we have to emphasize that they were all reported in the context of *transmitter-induced plasticity* (more precisely corresponding to the synaptic adaptation, a non-Hebbian plasticity form). In biology, the facilitating processes observed in short timescale (i.e., *transmitter-induced STP*) and associated with an increase of neurotransmitter release probability during a burst of spike (i.e., corresponding to an increase of synaptic efficiency at high-frequency spiking rate) is additive with LTP [6] that could be associated with a Hebbian-type plasticity involving both pre- and post-neuron activities. In other words, a causal description will make a clear distinction between the origin of ST- and LT-plasticity while a phenomenological description (Fig. 10) will not. Indeed, during high-frequency burst of spikes associated with *transmitter-induced plasticity*, the firing of the post-neuron is favored and should lead to both pre- and post-activities, thus leading to Hebbian-type LTP. In the case of the neuromorphic implementation described above, the transition between STP and LTP is associated with a single parameter (such as the mean firing rate of the pre-neuron) and both ST and LT regimes cannot be uncorrelated (i.e., ST will lead to LT regime). The device state will move sequentially from one regime to another one via *transmitter-induced plasticity* only. It should be noted that this effect induces some restriction in terms of (i) network configurability, since non-Hebbian and Hebbian-type learning cannot be dissociated, and (ii) network functionality, since the synaptic connection moves from a nonlinear conductance in its ST regime (i.e., frequency dependent) to a linear conductance in its LT regime. Alternative approaches are still needed as proposed by Cantley et al. [11] where short-term processes and long-term processes are realized by two different devices (leaky floating gate transistor and nonvolatile two-terminal devices) in order to match the complexity of biological synapses.

Another approach [23] relies on the fact that ECM cells are multi-filamentary systems providing one additional parameter for device's conductance modulation: Either the number of filaments or the size of a single filament can produce an increase of conductivity, while these two situations will lead to different volatility properties (Fig. 11). The independent control of these two parameters leading both to potentiation offers the possibility to dissociate different forms of plasticity and to reproduce synaptic plasticity in a more biorealistic way. In particular, in the case of multi-filamentary ECM cells, an independent control of the number of filaments and of the width of each individual filament was proposed in order to reproduce different potentiation with both ST and LT regimes.

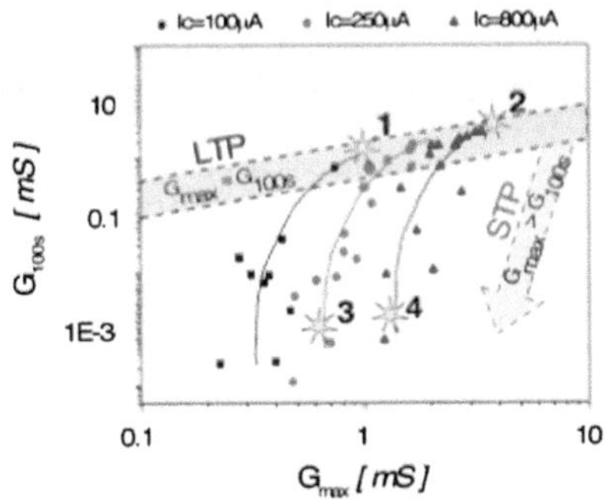

Fig. 11 STP and LTP implementation in an ECM cell. By using the number of pulses as plasticity key factor, two examples of LTP (case 1 and 2) and STP (case 3 and 4) are obtained. Both dendritic branches density and dendrites diameter can be tuned independently to reproduce various STP/LTP combinations

5 Conclusions

We have presented various plasticity mechanisms that have been implemented in nanoscale memory devices, promising candidates for future biomimetic hardware systems. Of course, the different examples described above are far from being exhaustive but are a tentative classification and formalization of synaptic plasticity in nanoscale devices. Future works should provide more complex device systems with richer features embedded in nanoscale components that will pave the way to complex neuromorphic computing systems. Notably, while we have only focused on the synaptic aspect, important efforts are still needed to implement neurons and synaptic interconnections that will determine the applicability of the different concepts exposed in this chapter. Indeed, since synaptic elements are required to be implemented in a high-density architecture, major challenges in terms of practical operating conditions and interconnections strategies should be taken into account.

Finally, neuromorphic computing being an emerging field evolving in between ANNs and BNNs, strong interdisciplinary approaches will be valuable for the future of neuromorphic computing.

Acknowledgements The authors thank Dr. Dominique Vuillaume for careful reading of the manuscript and Dr. Damien Querlioz for fruitful discussions. This work was supported by ANR-12-PDOC- 0027-01 (Grant DINAMO).

References

1. Abbott, L., Varela, J., Sen, K., Nelson, S.: Synaptic depression and cortical gain control. Science **275**(5297), 221–224 (1997)
2. Abbott, L.F., Nelson, S.B.: Synaptic plasticity: taming the beast. Nat. Neurosci. **3**, 1178–1183 (2000)

3. Alibart, F., Pleutin, S., Guérin, D., Novembre, C., Lenfant, S., Lmimouni, K., Gamrat, C., Vuillaume, D.: An organic nanoparticle transistor behaving as a biological spiking synapse. Adv. Funct. Mater. **20**(2), 330–337 (2010)
4. Bi, G.-Q., Poo, M.-M.: Synaptic modifications in cultured hippocampal neurons: dependence on spike timing, synaptic strength, and postsynaptic cell type. J. Neurosci. **18**(24), 10464–10472 (1998)
5. Bienenstock, E.L., Cooper, L.N., Munro, P.W.: Theory for the development of neuron selectivity: orientation specificity and binocular interaction in visual cortex. J. Neurosci. **2**(1), 32–48 (1982)
6. Bliss, T.V., Collingridge, G.L., et al.: A synaptic model of memory: long-term potentiation in the hippocampus. Nature **361**(6407), 31–39 (1993)
7. Boegerhausen, M., Suter, P., Liu, S.-C.: Modeling short-term synaptic depression in silicon. Neural Comput. **15**(2), 331–348 (2003)
8. Boyden, E.S., Katoh, A., Raymond, J.L.: Cerebellum-dependent learning: the role of multiple plasticity mechanisms. Neuroscience **27** (2004)
9. Buonomano, D.V., Karmarkar, U.R.: Book review: how do we tell time? Neurosc. **8**(1), 42–51 (2002)
10. Buonomano, D.V., Maass, W.: State-dependent computations: spatiotemporal processing in cortical networks. Nat. Rev. Neurosci. **10**(2), 113–125 (2009)
11. Cantley, K.D., Subramaniam, A., Stiegler, H.J., Chapman, R., Vogel, E.M., et al.: Hebbian learning in spiking neural networks with nanocrystalline silicon tfts and memristive synapses. IEEE Trans. Nanotechnol. **10**(5), 1066–1073 (2011)
12. Chang, S.H., Lee, S.B., Jeon, D.Y., Park, S.J., Kim, G.T., Yang, S.M., Chae, S.C., Yoo, H.K., Kang, B.S., Lee, M.-J., et al.: Oxide double-layer nanocrossbar for ultrahigh-density bipolar resistive memory. Adv. Mater. **23**(35), 4063–4067 (2011a)
13. Chang, T., Jo, S.-H., Kim, K.-H., Sheridan, P., Gaba, S., Lu, W.: Synaptic behaviors and modeling of a metal oxide memristive device. Appl. Phys. A **102**(4), 857–863 (2011b)
14. Clopath, C., Büsing, L., Vasilaki, E., Gerstner, W.: Connectivity reflects coding: a model of voltage-based stdp with homeostasis. Nat. Neurosci. **13**(3), 344–352 (2010)
15. Deng, Y., Josberger, E., Jin, J., Rousdari, A.F., Helms, B.A., Zhong, C., Anantram, M., Rolandi, M.: H+-type and oh–type biological protonic semiconductors and complementary devices. Sci. Rep. **3** (2013)
16. Desbief, S., Kyndiah, A., Guerin, D., Gentili, D., Murgia, M., Lenfant, S., Alibart, F., Cramer, T., Biscarini, F., Vuillaume, D.: Low voltage and time constant organic synapse-transistor. Org. Electron. **21**, 47–53 (2015)
17. Du, C., Ma, W., Chang, T., Sheridan, P., Lu, W.D.: Biorealistic implementation of synaptic functions with oxide memristors through internal ionic dynamics. Adv. Funct. Mater. **25**(27), 4290–4299 (2015)
18. Gjorgjieva, J., Clopath, C., Audet, J., Pfister, J.-P.: A triplet spike-timing-dependent plasticity model generalizes the bienenstock-cooper-munro rule to higher-order spatiotemporal correlations. Proc. Natl. Acad. Sci. **108**(48), 19383–19388 (2011)
19. Hebb, D.O.: The first stage of perception: growth of the assembly. Org. Behav. 60–78 (1949)
20. Izhikevich, E.M., et al.: Simple model of spiking neurons. IEEE Trans. Neural Netw. **14**(6), 1569–1572 (2003)
21. Kim, S., Du, C., Sheridan, P., Ma, W., Choi, S., Lu, W.D.: Experimental demonstration of a second-order memristor and its ability to biorealistically implement synaptic plasticity. Nano Lett. **15**(3), 2203–2211 (2015)
22. Kuzum, D., Jeyasingh, R.G., Lee, B., Wong, H.-S.P.: Nanoelectronic programmable synapses based on phase change materials for brain-inspired computing. Nano Lett. **12**(5), 2179–2186 (2011)
23. La Barbera, S., Vuillaume, D., Alibart, F.: Filamentary switching: synaptic plasticity through device volatility. ACS Nano **9**(1), 941–949 (2015)
24. Lamprecht, R., LeDoux, J.: Structural plasticity and memory. Nat. Rev. Neurosci. **5**(1), 45–54 (2004)

25. Lim, J., Ryu, S.Y., Kim, J., Jun, Y.: A study of tio2/carbon black composition as counter electrode materials for dye-sensitized solar cells. Nanoscale Res. Lett. **8**(1), 1–5 (2013)
26. Maass, W., Natschläger, T.: Networks of spiking neurons can emulate arbitrary hopfield nets in temporal coding. Netw. Comput. Neural Syst. **8**(4), 355–371 (1997)
27. Malenka, R.C., Bear, M.F.: Ltp and ltd: an embarrassment of riches. Neuron **44**(1), 5–21 (2004)
28. Markram, H., Lübke, J., Frotscher, M., Sakmann, B.: Regulation of synaptic efficacy by coincidence of postsynaptic aps and epsps. Science **275**(5297), 213–215 (1997)
29. Markram, H., Pikus, D., Gupta, A., Tsodyks, M.: Potential for multiple mechanisms, phenomena and algorithms for synaptic plasticity at single synapses. Neuropharmacology **37**(4), 489–500 (1998)
30. Mayr, C., Partzsch, J., Noack, M., Schüffny, R.: Live demonstration: multiple-timescale plasticity in a neuromorphic system. In: ISCAS, pp. 666–670 (2013)
31. Mayr, C., Stärke, P., Partzsch, J., Cederstroem, L., Schüffny, R., Shuai, Y., Du, N., Schmidt, H.: Waveform driven plasticity in bifeo3 memristive devices: model and implementation. In: Advances in Neural Information Processing Systems, pp. 1700–1708 (2012)
32. McCulloch, W.S., Pitts, W.: A logical calculus of the ideas immanent in nervous activity. Bull. Math. Biophys. **5**(4), 115–133 (1943)
33. Ohno, T., Hasegawa, T., Tsuruoka, T., Terabe, K., Gimzewski, J.K., Aono, M.: Short-term plasticity and long-term potentiation mimicked in single inorganic synapses. Nat. Mater. **10**(8), 591–595 (2011)
34. Senn, W., Markram, H., Tsodyks, M.: An algorithm for modifying neurotransmitter release probability based on pre-and postsynaptic spike timing. Neural Comput. **13**(1), 35–67 (2001)
35. Sjöström, P.J., Turrigiano, G.G., Nelson, S.B.: Rate, timing, and cooperativity jointly determine cortical synaptic plasticity. Neuron **32**(6), 1149–1164 (2001)
36. Snider, G.S.: Spike-timing-dependent learning in memristive nanodevices. In: IEEE International Symposium on Nanoscale Architectures, 2008. NANOARCH 2008, pp. 85–92. IEEE (2008)
37. Sourdet, V., Debanne, D.: The role of dendritic filtering in associative long-term synaptic plasticity. Learn. Mem. **6**(5), 422–447 (1999)
38. Strukov, D.B., Snider, G.S., Stewart, D.R., Williams, R.S.: The missing memristor found. Nature **453**(7191), 80–83 (2008)
39. Subramaniam, A., Cantley, K.D., Bersuker, G., Gilmer, D., Vogel, E.M.: Spike-timing-dependent plasticity using biologically realistic action potentials and low-temperature materials. IEEE Trans. Nanotechnol. **12**(3), 450–459 (2013)
40. Van Rossum, M.C., Bi, G.Q., Turrigiano, G.G.: Stable hebbian learning from spike timing-dependent plasticity. J. Neurosci. **20**(23), 8812–8821 (2000)
41. Varela, J.A., Sen, K., Gibson, J., Fost, J., Abbott, L., Nelson, S.B.: A quantitative description of short-term plasticity at excitatory synapses in layer 2/3 of rat primary visual cortex. J. Neurosci. **17**(20), 7926–7940 (1997)
42. Wang, Z.Q., Xu, H.Y., Li, X.H., Yu, H., Liu, Y.C., Zhu, X.J.: Synaptic learning and memory functions achieved using oxygen ion migration/diffusion in an amorphous ingazno memristor. Adv. Funct. Mater. **22**(13), 2759–2765 (2012)
43. Williamson, A., Schumann, L., Hiller, L., Klefenz, F., Hoerselmann, I., Husar, P., Schober, A.: Synaptic behavior and stdp of asymmetric nanoscale memristors in biohybrid systems. Nanoscale **5**(16), 7297–7303 (2013)
44. Yang, Y., Choi, S., Lu, W.: Oxide heterostructure resistive memory. Nano Lett. **13**(6), 2908–2915 (2013)
45. Yuan, P., Leonetti, M.D., Pico, A.R., Hsiung, Y., MacKinnon, R.: Structure of the human bk channel ca2+-activation apparatus at 3.0 å resolution. Science **329**(5988), 182–186 (2010)
46. Zenke, F., Agnes, E.J., Gerstner, W.: Diverse synaptic plasticity mechanisms orchestrated to form and retrieve memories in spiking neural networks. Nat. Commun. **6** (2015)
47. Ziegler, L., Zenke, F., Kastner, D.B., Gerstner, W.: Synaptic consolidation: from synapses to behavioral modeling. J. Neurosci. **35**(3), 1319–1334 (2015)

Neuromemristive Systems: A Circuit Design Perspective

Cory Merkel and Dhireesha Kudithipudi

Abstract Neuromemristive systems (NMSs) are brain inspired, adaptive computer architectures based on emerging resistive memory technology (memristors). NMSs adopt a mixed-signal design approach with closely coupled memory and processing, resulting in high area and energy efficiencies. Existing work suggests that NMSs could even supplant conventional architectures in niche application domains. However, given the infancy of the field, there are still a number open design questions, particularly in the area of circuit realization, that must be explored in order for the research to move forward. This chapter reviews a number of theoretical and practical concepts related to NMS circuit design, with particular focus on neuron, synapse, and plasticity circuits.

1 Introduction: Taking a Cue from Nature

An NMS is a brain inspired, special-purpose computing platform based on nanoscale resistive memory (memristor) technology. NMSs represent a subclass of a broader movement in brain-inspired computing called neuromorphic systems, which were pioneered by Carver Mead in the late 1980s [1]. The primary goal of both neuromorphic and neuromemristive systems is to provide levels of intelligent information processing, adaptation/learning, energy/area efficiency, and noise/fault tolerance in niche application domains that are not achievable using conventional computing paradigms. Conventional computer architectures are limited in these aspects because of their adherence to the von Neumann model, where the hardware is digital and immutable, computation is sequential and precise, and a distinct separation exists between computation and memory. Although the von Neumann model is

C. Merkel · D. Kudithipudi (✉)
Department of Computer Engineering, Rochester Institute of Technology,
1 Lomb Memorial Drive, Rochester, NY 14623-5603, USA
e-mail: dxkeec@rit.edu

C. Merkel
e-mail: c.merkel87@gmail.com

© Springer (India) Pvt. Ltd. 2017

M. Suri (ed.), *Advances in Neuromorphic Hardware Exploiting Emerging Nanoscale Devices*, Cognitive Systems Monographs 31,
DOI 10.1007/978-81-322-3703-7_3

unparalleled for well-defined sequential problems (e.g., arithmetic and logic), it is ill suited in application domains such as visual information processing, where problems are not well posed, data are analog and noisy, and solutions are inherently parallel. Mead and many researchers before him recognized that biological systems such as the primate brain solve these types of problems with much greater efficiency than conventional computing systems. In fact, it is estimated that for applications such as visual information processing, the brain is a factor 1×10^7 more energy efficient than any conceivable digital computer. The explanation for this large efficiency gap lies in the stark contrast between conventional computer architectures and the computing methods employed by the brain.

The human brain is inherently mixed signal, massively parallel, approximate, and plastic, giving rise to its incredible processing ability, low power consumption, and capacity for adaptation. Both neuromorphic and neuromemristive systems attempt to emulate brain functionality with neural networks built from mixed-signal circuits. The two distinguishing features of an NMS are as follows:

- The incorporation of memristive devices into NMSs enables plasticity at multiple levels, beyond the synaptic plasticity that is typically implemented in neuromorphic systems.
- NMSs focus on abstraction of the computational principles found in the nervous system rather than biological plausibility. This approach is better for two reasons. First, it is still unclear how behavior at the level of single neurons and small neuronal populations leads to system-level behavior of the brain. Secondly, the basic components of the brain (e.g., proteins and cells) are much different than those used in integrated circuit (IC) design (e.g., transistors and memristors). Therefore, it is unlikely that copying the brain's structure in an IC will yield the same emergent properties.

Note that there are other computing platforms that are attractive for brain-like information processing, including general-purpose graphical processing units (GPGPUs) and field-programmable gate arrays (FPGAs). GPGPUs are optimized for the types of linear algebra computations that govern neural network behavior. However, they lack the reconfigurability that is offered by NMSs. On the other hand, FPGAs have a high degree of reconfigurability, but they have very high area and power overheads to support their interface and routing resources.

NMSs generally have neural network-like architectures/topologies. To date, a number of NMS designs have been proposed for a wide spectrum of applications. These include associative memory [2], brain-state-in-a-box recall [3], temporal pattern recognition in a reservoir network [4], and implication logic [5]. Beyond these applications, many groups have focused on NMSs for vision. In [6], a neuromemristive winner-take-all-type network is designed to detect the position of an object. A type of biologically motivated training called spike time-dependent plasticity (STDP) is used for unsupervised learning. The authors of [7] propose an NMS with a multi-layer perceptron topology for classifying automobiles. An NMS for optical character recognition (OCR) is designed in [8] using a simple feedforward network and STDP training. In [9], the authors propose an NMS that uses stochastic conductive bridge random access memory (CBRAM) devices for visual pattern extraction, and in [10] an NMS with an extreme learning machine topology is designed for pattern recognition.

The rest of this chapter focuses on the circuit designs underlying the architectures cited above. While the emphasis will be on circuit functionality, models of the circuits' power consumption, area, and variability can be found in the accompanying references. Section 2 provides a brief overview of memristors and the motivation for their incorporation into NMSs. Section 3 discusses voltage versus current-mode circuit designs for NMSs. Sections 4, 5, and 6 review circuit designs for neuron, synapse, and plasticity (learning) circuits, respectively. Section 7 concludes this chapter.

2 Memristor Overview

Our abilities to learn a new face, drive a car, identify objects, and perform other visual tasks are the results of brain plasticity. Plasticity is the characteristic of a system that allows it to undergo permanent changes in response to an external force. Biological systems exhibit remarkable levels of plasticity, enabling organisms to adapt to a changing environment, maintain a homeostatic state, and recover from injury. The same characteristics are of interest for future computing systems as they facilitate reconfigurability and noise tolerance, reliability, and self-healing/resilience. The mechanisms that enable plasticity in a biological context occur at multiple scales, from the level of individual cells up to functional brain regions. These include neurogenesis, epigenetic mechanisms, long-term potentiation and depression in chemical synapses, and changes in topological mappings between brain regions and brain functions (e.g., retinotopic maps). At an abstract level, each of these plasticity mechanisms requires some form of memory. In particular, there is a certain level of persistence in, e.g., the locations of specific neurons, the efficacy of synaptic transmission in a particular synapse, and the topology of brain regions. Hence, any brain-inspired computing system should ideally employ some form of nonvolatile memory (NVM) to achieve plastic behavior.

Flash has been the dominant nonvolatile memory technology used in computing systems for many years because of its high density and low cost. However, due to many scaling-related challenges, flash is expected to be superseded by a novel memory technology within the next decade. Table 1 shows a comparison of NAND flash and prototypical/emerging nonvolatile memories across energy, performance, and reliability metrics. Biologically motivated targets for each metric are listed in the right column. In particular, phase change memory (PCM), spin transfer torque random access memory (STT-RAM), and resistive random access memory (RRAM) are among the most promising candidates for future NVM implementations [11]. Each of these technologies may also be described as a memristor or memristive device. A memristor is a two-terminal passive circuit element that follows a state-dependent Ohm's Law, characterized by a pinched hysteresis current–voltage relationship as shown in Fig. 1 [12–14]:

Table 1 Comparison of nonvolatile memory technologies for brain-inspired computing [11, 15–17]

Metric	Flash	Memristors			Targets
		PCM	STT-RAM	RRAM	
Dynamic range (℧/℧)	–	>1000	2	1000	>4
Number of states	8–16	100	4	100	20–100
Retention	Several years at room temp.				Years
Energy (pJ/bit)	>100	2–25	0.1–2.5	0.1–3	0.01
Endurance (cycles)	10^4	10^9	10^{15}	10^{12}	10^9

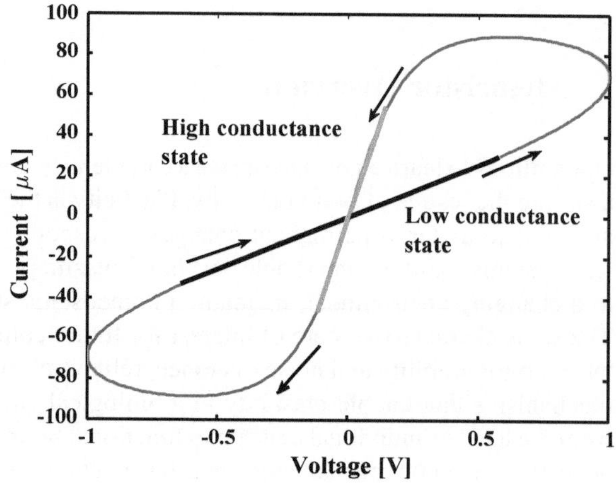

Fig. 1 Pinched hysteresis current–voltage relationship that characterizes memristive devices

$$i_m(t) = G_m(\gamma)v_m(t) \tag{1}$$

$$\frac{\mathrm{d}\gamma}{\mathrm{d}t} = \chi(\gamma, v_m(t)) \tag{2}$$

where i_m is the current through the memristor, v_m is the voltage across the memristor, $\gamma \in [0, 1]$ is a state variable, $G_m(\gamma)$ is the state-dependent conductance, and χ governs how γ changes over time. The conductance will range from $G_{moff} \equiv G_m(\gamma = 0)$ to $G_{mon} \equiv G_m(\gamma = 1)$. By applying short voltage pulses to these devices, one can incrementally modify their conductance states, enabling the storage of multi-level memory values.

The most important metrics for an NVM technology within an NMS are dynamic range, number of memory states, retention, energy efficiency, and endurance. A memristor's dynamic range can be measured as the ratio of its *ON* and *OFF* conductances (G_{mON}/G_{mOFF}). A large dynamic range allows sense circuitry to easily distinguish an NVM cell's different memory states. The number of states that the NVM cell can achieve has a direct impact on the area and energy efficiencies, as well as the functionality of an NMS. The number of memory states in a memristive device is equivalent to the number of distinguishable $G_m(\gamma)$ values that exist. For two conductance states to be distinguishable, they need to yield two different current levels (when placed in a circuit) that have a range which is larger than the noise level (e.g., thermal and shot noise) of the circuit. It may take several bistable (only able to achieve two memory states) NVM cells in an NMS to attain the same level of functionality as an NMS with a single NVM cell that has many memory states. Retention is another critical characteristic for an NVM technology. Within an NMS, a large retention allows the system to accumulate and integrate information over long periods of time. Low power is a primary NMS design goal, making energy-per-bit a critical metric in evaluating NVM technologies within these systems. Finally, in order for an NMS to learn and adapt, its underlying memory must be able to endure a large number of write events. Based on these metrics, RRAM is the most suitable NVM for NMS implementation. Although it has a good dynamic range, number of states, and retention, PCM requires high energy and voltages for writing. In addition, its endurance is borderline. In contrast, STT-RAM has very high endurance, but its dynamic range and number of resistance states are too small for NMS implementation. In addition, RRAM has excellent compatibility with CMOS and is highly scalable; its competition with other emerging NVM technologies will continue to fuel research that will be fruitful for RRAM-based NMSs.

RRAM cells, which will be referred to as memristors for the rest of this document, have a metal–insulator–metal (MIM) structure, where two conducting electrodes sandwich a thin-film switching layer. Various MIM memristor stacks have been explored, and there are several ways to categorize them based on their material properties (e.g., crystalline structure and band gap), proposed switching mechanism (e.g., anion, cation, and Ferroelectric), or observed switching characteristics (e.g., bipolar or unipolar switching). Comprehensive reviews are provided in [15, 18–21]. The proposed switching mechanism for most of the fabricated devices is based on redox reactions and migration of defects such as interstitial ions or vacancies. Several different models have been proposed to capture the physical phenomena underlying memristive behavior [21–24]. However, many of them are computationally expensive and are not amenable to large-scale circuit simulations. Simpler empirical models such as the PWL model proposed in [25] are parametrized by experimental memristor data and have lower computational complexity. Finally, several groups have proposed simulation program with integrated circuit emphasis (SPICE) or Verilog AMS models for circuit-level simulations [26–33].

3 Voltage Versus Current-Mode Circuit Designs for NMSs

At the circuit level, an NMS can perform computations on information represented by currents (current mode), voltages (voltage mode), or a combination of the two. Deciding which of these methodologies to adopt can be challenging, as each has its own set of strengths and weaknesses. Current-mode design techniques date back to 1975, when Gilbert proposed a general class of circuits composed of devices (called translinear elements) that have an exponential current–voltage relationship [34]. Gilbert's *translinear principle* states that circuits configured with loops of translinear elements (translinear circuits) behave in a very predictable way: The products of the currents flowing in one direction equal the products of the currents flowing in the other direction. Initially, the translinear principle was demonstrated with bipolar junction transistor (BJT) devices, but it is also applicable to MOS devices operating in weak inversion. Complex computations such as vector magnitude calculations can be implemented in current mode using the translinear principle with a handful of transistors. There is no similar design principle for voltage-mode circuits.

Current-mode circuits have several other advantages over voltage-mode designs. They are generally able to operate at lower supply voltages and typically can achieve higher bandwidths, sometimes approaching the MOSFET intrinsic frequency f_T [35]. In addition, current representations of information have an inherent advantage in terms of communication. When voltages are sent along long routing paths, they incur losses due to series resistances, diminishing the integrity of the signal. Biology's solution to this problem has been to send long-range communications in the form of spikes which are regenerated along myelinated axons. However, it is still largely unknown how neural information is encoded in spikes and rate encoding is still the dominant scheme used in hardware implementations of spiking networks. It is often easier to represent spiking rates in hardware as continuous analog values, albeit with some reduced noise tolerance. However, buffering analog voltages requires expensive hardware, with carefully designed amplifier circuits (e.g., common drain amplifiers) to achieve unity gain. In addition, simple analog voltage buffers typically operate in small-signal operating regions and require very careful biasing to obtain zero offset. Better designs typically employ a source follower op-amp configuration which can handle rail-to-rail input and output signals. However, even the simplest op-amps consisting of differential and gain stages require 7 transistors—contrast that with current-mode designs, which can communicate information over long distances with relatively little signal degradation.

In addition to the general advantages of current-mode circuits, there are also specific benefits when these designs are used to implement neuromemristive architectures and systems. Consider the two configurations in Fig. 2. In the first case (Fig. 2a), a presynaptic neuron has a voltage output $v_{x_j^{(l)}}$ which falls across an output impedance R_o connected to a small-signal ground. Here, each neuron is modeled as an ideal voltage source that implements a linear activation function. However, the pre- and postsynaptic neurons may generally have any activation function. The gain of the presynaptic neuron can be described as $A_v = g_m R_o$, where g_m is a transconductance factor. Now, consider the direct connection of the presynaptic neuron to all of the outgoing synapses, which can be characterized by several parallel conductance values. In this case, the gain of the presynaptic neuron becomes $A_v = g_m \left(R_o || \cdots \frac{1}{G_{i-1,j}^{(l+1)}} || \frac{1}{G_{i,j}^{(l+1)}} || \frac{1}{G_{i+1,j}^{(l+1)}} \cdots \right)$. Therefore, if the neuron's fan-out is high, then small weight values in the outgoing synapses (typically represented as high conductances) will significantly diminish the neuron gain. Therefore, it is usually best

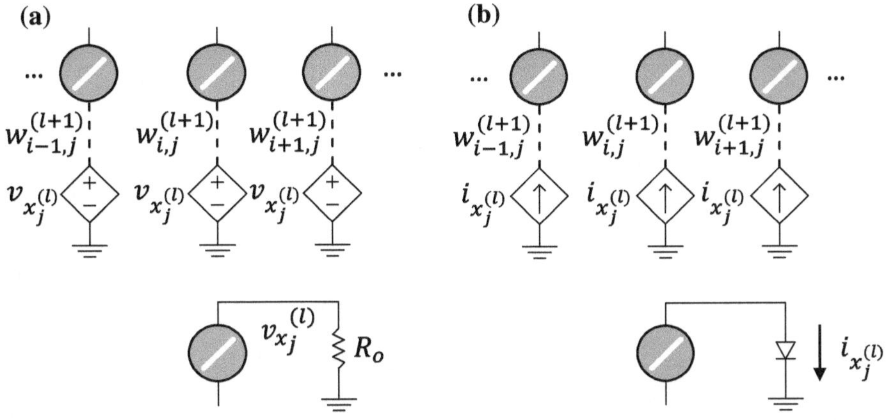

Fig. 2 **a** Voltage-mode NMS, where neuronal activations are represented as voltages and synapses typically operate via transconductance. **b** Current-mode NMS where neuronal activations and the results of synaptic weighting are represented as currents

practice to add a buffer before each outgoing synapse to reduce the loading effect on the presynaptic neuron. In fact, this technique is analogous to biological nervous systems, where each outgoing synapse is buffered using synaptic vesicles held within the presynaptic terminal (synaptic bouton). As discussed earlier, buffering voltage-mode analog neurons is expensive in terms of hardware area. In contrast, a current-mode design (Fig. 2b) affords the ability to buffer presynaptic neuron outputs using current-controlled current sources (typically simple CMOS current mirrors), which have smaller area and power requirements.

Although the current-mode design approach is attractive, there are some challenges to consider when designing NMSs using current-mode circuits. First, a current cannot be distributed through multiple circuit branches without buffering. Secondly, since current mirrors are employed extensively, current-mode designs are especially prone to mismatch-related problems. The effects of mismatch on the circuit, architecture, and system-level performance are studied extensively in this work.

4 Neuron Circuits: Primary Information Processing Units

The human brain contains an estimated 80 billion neuron cells. In an abstract sense, each neuron performs the same task; it processes information sent from one group of neurons and then sends the result along to another group of neurons. Of course, this is a massive oversimplification. In reality, there are hundreds of types of neurons, each having complex and unique physical, biochemical, and functional characteristics.[1]

[1] For example, different neurons may have different dendritic arborizations, axon lengths, methods of encoding/decoding information, etc.

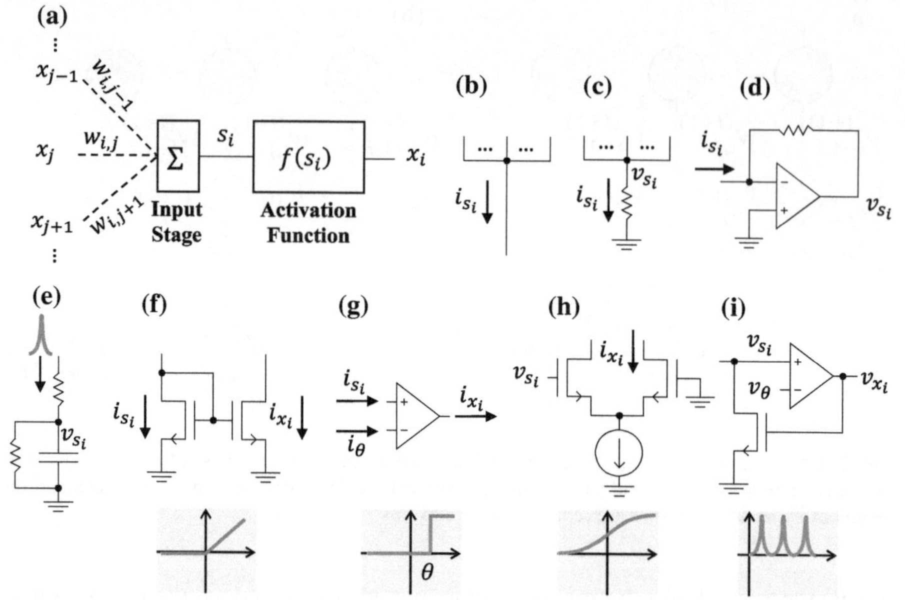

Fig. 3 NMS neuron circuit designs. **a** Block diagram of a neuron showing input and activation function stages. **b–d** Input stages using a common node for current summation, a pull-down resistor for current-to-voltage conversion, and an inverting summing amplifier circuit to provide a virtual ground node, respectively. **e–h** Example circuits for implementing rectified linear, threshold, sigmoid, and spiking activation functions, respectively

Although these nuances are critical from a biological standpoint, it is infeasible to capture so many details in an energy-efficient/area-efficient circuit. Therefore, NMSs typically employ simplified behavioral models of neurons, such as those shown in Fig. 3.

In the first stage, or input stage (see Fig. 3a), each neuron sums weighted outputs from other neurons. This convergence of signals allows each neuron to integrate information from across the network and get its own unique sense of the system's current state. In the second stage, the neuron applies an activation function to the summation and sends the result downstream to other neurons that are awaiting its arrival. It is in this stage that information is said to diverge, since the downstream neurons may be distributed across the entire network. Overall, the behavior of a neuron in an NMS can be described as

$$x_i(t) = f(s_i(t)) = f\left(\sum_j s_{i,j}(t)\right), \tag{3}$$

where f is the activation function and $s_{i,j}$ is the jth incoming signal to the ith neuron.

4.1 Input Stage

The neuron input stage defines how signals are integrated over space/time. Depending on the design approach, the input stage will place different constraints on the neuron, such as its maximum fan-in. Figure 3b–3e show four common input stages. The simplest design (Fig. 3b, c) makes use of Kirchoff's current law, connecting all of the incoming signals, which are assumed to be currents, to a common node for summation. Optionally, a pull-up/pull-down resistor may be used to convert the current summation to a voltage. The choice of whether or not to include a current-to-voltage conversion depends on the design of the activation function circuit, discussed later. This type of input stage (with and without voltage conversion) has generally been used in NMSs with so-called first- and second-generation designs [10, 36–38], where neurons are nonspiking and transient states of the network are ignored (except in the case of recurrent topologies). In the case where a current-to-voltage conversion is required, it is advisable to use a memristor in place of a pull-up/pull-down resistor. The reason is twofold: First, the currents that appear at the neuron's input stage are usually in the nanoampere range, so a very large resistance is needed to get a good dynamic voltage range. Large resistors are very costly (in terms of area) to implement on-chip. Secondly, the ability to adjust a memristor's resistance/conductance can be useful for adjusting the neuron's activation function.

Neuron input stages can also be designed using operational amplifiers, as shown in Fig. 3d. The advantage of this approach is that the amplifier circuits can be used to force identical potentials at different circuit nodes, which is especially useful when designing NMSs that incorporate memristor crossbars [10, 39]. The disadvantage, which is frequently overlooked, is that good op-amps (e.g., meeting minimum phase margin, slew rate, gain, and input/output range) require significant effort from experienced analog designers, not to mention their potentially large area and power costs. In fact, the open-loop gain is usually assumed to be infinite [39], allowing the circuit to produce a perfect summation even with a very large fan-in. However, in [10], the authors show that op-amp input stages have very limited fan-in (e.g., $\approx$50 incoming synapses), even when the op-amp gain is high (e.g., 100 dB). However, this constraint can be relaxed if it is not necessary for the op-amp circuit to have perfect behavior (i.e., produce a perfect summation of the inputs). Similar to the input stage discussed earlier, an op-amp-based input stage may also need to have very large resistances, so a memristor implementation is much more practical than, e.g., a polysilicon resistor implementation.

The input stages discussed so far integrate incoming signals over space, and the networks that they comprise make use of steady-state signals (again, with the exception of recurrent networks). Another type of input stage integrates signals over space and time, allowing an NMS to exploit temporal dynamics for efficient encoding and communication of information. The circuit is shown in Fig. 3e. The biological motivation is readily apparent. An input resistor and a capacitor represent the impedance and capacitance of the neuron cell's membrane, respectively. Charge is integrated on the capacitor each time a voltage spike, or action potential, appears at the input. The current flowing into the capacitor represents sodium influx that occurs after a neuron receives an incoming action potential. The current flowing out of the capacitor represents potassium efflux, which causes the charge to "leak," pulling the membrane voltage down to the neuron's resting potential. Consequently, the neuron model associated with the input stage in Fig. 3e is referred to as a leaky-integrate-and-fire (LIF) model. The LIF input stage can be designed in a variety of ways. For example, the current injected into the capacitor after receiving an action potential and the leakage current can both be made approximately independent of the membrane potential itself. See [40] for a review.

4.2 Activation Function

Possibly the most critical aspect of a neuron in an NMS (and any other artificial neural network) is its activation function f. The precise shape of f dictates the complexity of relationships that the NMS can learn. Recall that real (biological) neurons encode information in spikes. Generally, a neuron's spike frequency will increase as it becomes more excited and vice versa. Perhaps the simplest way to model this behavior is with an identity function. In other words, the output of the activation function will be the same as its input. Of course, this type of activation function does not need to be implemented explicitly, since one can just take the output as s_i. But if all of the activation functions in the network are linear, then it follows that the NMS will only be able to learn linear relationships. Fortunately, the linear activation function can be made nonlinear by adding a diode-connected MOSFET, as shown in Fig. 3f. This type of activation function is referred to as a rectified linear unit (ReLU). Besides its computational advantages, ReLUs are much more biologically plausible than linear activation functions. Despite their simple circuit implementation, ReLUs have scarcely been incorporated into NMS designs. However, that is likely to change given the ReLU's success in software implementations of deep belief networks.

Another simple activation function models a neuron as being either "ON" or "OFF." Concretely, the circuit uses a binary signal, which is "1" to indicate a spike frequency above some threshold θ and "0" otherwise. The circuit implementation (Fig. 3g) consists of a comparator to implement the thresholding and a buffer to digitize the output. This type of threshold activation function is used in single-layer perceptron networks, which were proposed by Rosenblatt in the late 1950s [41]. Although Minsky and Papert showed in the following decade that single-layer perceptrons have a major flaw, they can only learn linearly separable functions [42]. To learn more complex relationships, an NMS must employ multilayer networks with continuous and nonlinear activation functions, such as logistic sigmoid and hyperbolic tangent functions. The circuit implementation of a logistic sigmoid function is shown in Fig. 3h. Note that if the output of the circuit is taken as the difference of the two differential pair branch currents, then the result will be a hyperbolic tangent function. Logistic sigmoid and hyperbolic tangent functions are very similar, and their "S" shape is often used to model the spike rates of biological neurons [43].

Both functions have been used in the implementation of NMSs for a variety of applications [10, 39, 44]. The major difference between the two lies in their range. A logistic sigmoid function ranges from 0 to +1, while a hyperbolic tangent function ranges from −1 to +1. It has been shown empirically that bipolar (i.e., ranging from negative to positive values) activation functions perform better than unipolar activation functions (strictly positive or negative range) [45]. However, it can be shown that any network with bipolar activation functions can be transformed to one with unipolar activation functions [46]. Therefore, it is generally better to use unipolar activation functions because of their simpler circuit implementation.

The reason that Rosenblatt's perceptron could only learn linearly separable functions was not because it had a single layer. Rather, it was because the network's activation functions (i.e., the threshold activation function) were monotonic. In fact, all of the activation functions discussed so far are monotonic, meaning that they have a purely positive or purely negative first derivative. It can be shown that, by making the activation function nonmonotonic, a neuron can learn nonlinearly separable functions. A classic example of a nonlinearly separable function is parity detection. Another example is edge detection in images. In [47], researchers show that an NMS employing a nonmonotonic activation function can learn to detect edges using a single neuron. Nonmonotonic activation functions can be implemented as a combination of sigmoid activation circuits [47]. Although neurons in the brain exhibit nonmonotonic responses to external stimuli (e.g., cells in the primary visual cortex respond selectively to edge orientation), there is no evidence that their individual activation or encoding scheme is nonmonotonic. Therefore, an NMS neuron with a nonmonotonic activation function is modeling multiple layers of neurons in a biological neural network.

Evolution designed the neurons in our brains to encode information in spikes.[2] One can speculate that the evolutionary advantage of spiking is a mixture of efficiency and reliability. The former is evident from the brain's use of sparsely distributed encoding [48], and the latter can be inferred from neural tissue's leaky conduction properties, which would make the use of graded potentials very unreliable [43]. A number of NMS researchers have integrated spiking activation functions, such as the one shown in Fig. 3i, into their designs [9, 49]. One of the major questions concerning spiking NMSs is how should information be encoded/decoded? In nonspiking designs, especially feedforward networks, information is encoded by a neuron's intensity or the output of its activation function. However, when it comes to spiking neurons, there are several choices. Information can be encoded by spike rate, interspike latency, etc. [50]. To date, there are no studies that make a direct comparison between spiking and nonspiking NMSs in terms of area and energy efficiency, or computational capacity.

[2]However, photoreceptors in our eyes and neurons in the peripheral nervous system have graded (nonspiking) responses to stimuli.

5 Synapse Circuits: Communication and Memory

The neurons in our brains are functionally connected to each other via synapses. A biological synapse facilitates the transmission of information from one neuron to another using molecules called neurotransmitters. Neurotransmitters are released from special processes on the axon of one neuron (the presynaptic neuron) and diffuse across a small part of extracellular space to receptors on the dendrites of another neuron (the postsynaptic neuron). At the postsynaptic neuron, the neurotransmitters cause a change in concentration of intra- and extracellular ions (e.g., Na^{+}), leading to a change in the membrane potential. If the change in the membrane potential is positive and large enough, it could cause the postsynaptic neuron to spike. Some synapses may have a stronger effect (i.e., cause a larger change in membrane potential) on the postsynaptic neuron than others. Critically, the whole process of synaptic communication is plastic over different timescales. This means that the strength of communication between two neurons can change over time through a number of different processes including modulation of receptor efficacy, addition/deletion of receptors, synaptogenesis, and many others [51]. Synaptic plasticity is believed to be one of the primary mechanisms involved in learning.

The three primary functions of biological synapses (physical coupling of neurons, weighted communication, and facilitation of learning) can all be realized with memristor circuits. In the simplest design, often used in spiking NMSs, a single device connects pre- and postsynaptic neurons (Fig. 4a). This is often referred to as a 1R configuration. The device's conductance is modified to change the synaptic weight. The advantage of this approach is that it is the most compact. It also lends itself to a biologically motivated learning rule called spike time-dependent plasticity (STDP), which is discussed in the next section. However, there are a number of challenges related to a single-memristor design. Many devices have very low ON resistances (e.g., in 100 s of kΩ), causing adverse loading effects on presynaptic neurons, limiting their fan-out and drive capability. The design can be modified slightly (Fig. 4b) by adding a MOSFET to alleviate the loading effect. A variation of this design, which separates the presynaptic and postsynaptic neurons, but does not necessarily solve the loading issue is shown in Fig. 4c. This design has also been used in nonspiking designs [36].

A disadvantage of the designs discussed so far is that they can only achieve unipolar weight values, while artificial neural networks usually learn best with bipolar weights. The predominant method to get bipolar weight values has been to use two or more memristors per synapse with competing positive and negative effects on the postsynaptic neuron. One such design, proposed by Kim et al. [38], is shown in Fig. 4d. The circuit uses a bridge configuration to control the output of a differential pair. The amount of current flowing through each branch of the output depends on the ratios of memristor conductances. The final synaptic output is taken as the difference of the two branch currents, so the circuit can achieve both positive and negative weight values. In addition, the circuit has a reduced loading effect on the presynaptic neuron since the input impedance is approximately constant ($\approx (G_{on} + G_{off})/2$)

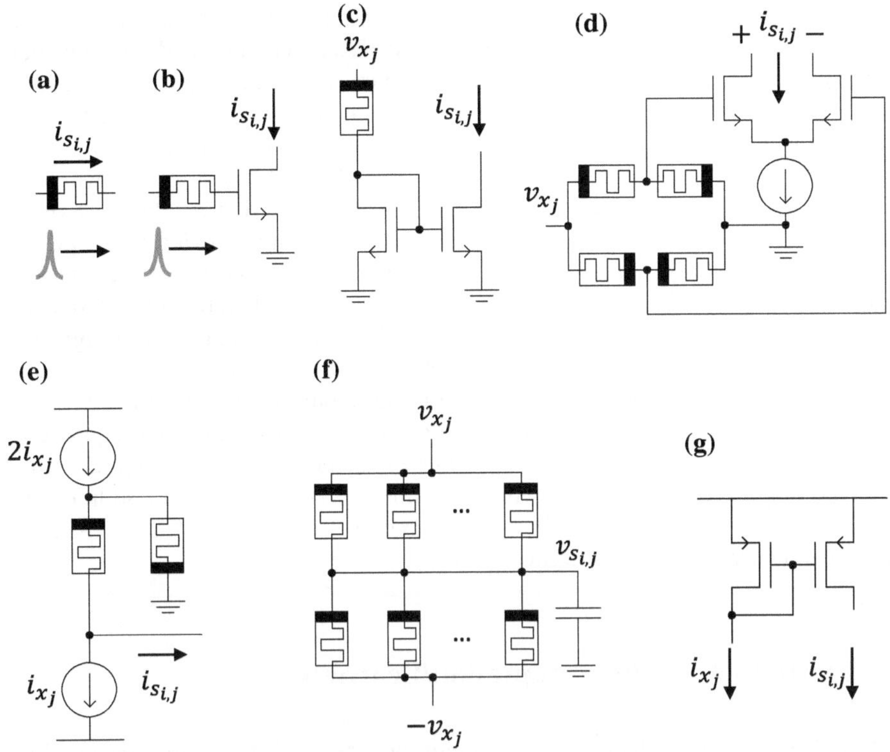

Fig. 4 Synapse circuits for NMSs: **a** Single memristor as a synapse, predominantly used in spiking networks. **b** Single-memristor synapse with a transistor used to provide a high-impedance input. **c** Single-memristor synapse with a current mirror to isolate the pre- and postsynaptic neurons. **d** Memristor bridge synapse providing bipolar weight values. **e** Current-mode bipolar weight synapse circuit. **f** Voltage-mode synapse composed of bistable memristors. **g** Constant weight synapse

over the entire weight range. Besides the large number of circuit components, one disadvantage of this design is that it only behaves linearly over a certain range of input voltages (i.e., when all transistors are saturated).

Another synapse design, shown in Fig. 4e, uses two memristors whose conductance ratio determines the synaptic weight. When the ratio is at one extreme, all of the current from the source current is shunted to ground, and the sink current inhibits the postsynaptic neuron. When the ratio is at the other extreme, the sink current is outweighed by the source current. Therefore, the circuit can achieve a continuous number of weight values between -1 and $+1$. Notice, however, that an op-amp is required to provide a virtual ground at the output node. As discussed earlier, op-amps are potentially expensive in terms of area, power consumption, and design effort.

It is important to note that not all memristors have continuous switching properties. That is, some devices can only achieve a small number of conductance states—the minimum of which is two. Devices with only two conductance states are referred to as bistable. Connecting a number N of bistable devices in parallel yields an equivalent memristor that has $N + 1$ states. Note that any of the synapse circuits discussed so far can be implemented with bistable devices. Each memristor would just be replaced with N bistable memristors in parallel. Figure 4f shows a bistable memristor-based synapse that incorporates an excitatory pull-up and an inhibitory pull-down group of memristors. Initially, the capacitor is discharged to ground. Then the positive and negative voltage representation of the presynaptic neuron is applied to the pull-up and pull-down networks, respectively. After a short time, the voltages are removed, and the capacitor will have charged toward v_{x_j} or $-v_{x_j}$ based on the relative states of the pull-up and pull-down conductances. The obvious disadvantage of designs that use bistable devices is the overhead associated with extra memristors.

One final synapse design is shown in Fig. 4g. Note the lack of any memristive devices in the circuit. This design is useful for network topologies that require constant, random weights, such as reservoir networks and extreme learning machines. The ratio of the transistor sizes will determine the weight value. In addition, for small transistor sizes, there will be some degree of random mismatch which can be leveraged to create random weight values.

6 Plasticity Circuits: Adaptation/Learning

NMSs learn through adaptation of synaptic weights which, in turn, are modified through adjustment of memristor conductance values. This section discusses a number of methods for adjusting memristor conductances in accordance with different learning algorithms. But before a learning algorithm can be developed, one must identify a cost function J. The cost function specifies how well the current state of the NMS (i.e., all of its weight values) satisfies the objective or hypothesis function (i.e., the desired functionality of the NMS). Once J is determined, it is straightforward to design a training algorithm that minimizes the cost via gradient descent:

$$w_{i,j} := w_{i,j} - \alpha \frac{\partial}{\partial w_{i,j}} J(\mathbf{W}), \tag{4}$$

where α is a constant called the learning rate, and $\mathbf{W}$ is the matrix containing all of the weights in the NMS. Now, consider a single-layer perceptron, which is the smallest indivisible unit of any neural network. If the current input is $\mathbf{u}^{(p)}$, the current output is $\hat{y}_i^{(p)}$, and the desired output is $y_i^{(p)}$, then it is easy to show that (4) becomes

$$w_{i,j} := w_{i,j} + \alpha u_j^{(p)} \left(y_i^{(p)} - \hat{y}_i^{(p)} \right) \tag{5}$$

This is the well-known least-mean-squares (LMS) algorithm. Here, p is the index of the current input. In general, increasing the weight value corresponds to applying a positive write voltage v_w to the synapse's memristors and vice versa. If the NMS is digital, all of the signals in (5) will be binary, and the weight update rule can be implemented using a digital circuit. On the other hand, if the NMS is analog/mixed signal, then implementation of (5) necessitates the use of an analog multiplier (Fig. 5a), which is expensive in terms of area, power, and design complexity.

A few methods have been developed to implement learning rules similar to (5) without the need for analog multiplication. In [52], the authors propose an adjusted learning rule:

$$w_{i,j}^{(p)} := w_{i,j}^{(p)} + \alpha \left(Y_i^{(p)} - \hat{Y}_i^{(p)} \right) U_j^{(p)}. \tag{6}$$

Uppercase letters correspond to Bernoulli random variables. This stochastic implementation of the least-mean-squares algorithm (SLMS) probabilistically increments and decrements weight values using only comparators and digital logic gates. Consequently, the circuit implementation consumes $\approx 3.5\times$ less area than the LMS implementation (Fig. 5b).

Another approach to training an NMS, shown in Fig. 6, is to approximate complex functions such as multiplication with functions that are easier to implement in hardware. The proposed circuit first converts an input current to a pulse width (left schematic). Then, using two current-to-pulse width (itop) converters and an AND gate, one can compute the min. In this case, the pulse width of `we` will be proportional to $\min(i_{x_j}, |i_{x_i} - i_{\hat{x}_i}|)$. The resulting write voltage is v_w. If the normalized values of $i_{\hat{x}_j}$ and $|i_{x_i} - i_{\hat{x}_i}|$ are both in the unit interval, then the normalized write voltage v_w will be very similar to their product.

The simplification presented above is used to design a learning circuit similar to the LMS rule [53], which can be written as

$$w_{ij} := w_{ij} + \alpha x_j x_{iD} = w_{ij} + \alpha x_j (\hat{x}_i - x_i). \tag{7}$$

Here, x_{iD} is the difference between the neuron's actual and expected outputs. In this work, a novel circuit for implementing a learning rule similar to (7). The modified training rule becomes

$$w_{ij} := w_{ij} + \alpha \mathrm{sign}(x_{iD}) \min(x_j, |x_{iD}|). \tag{8}$$

To implement this in hardware, one must first find $|x_{iD}|$ using a modified current subtraction circuit. Then, $\mathrm{sign}(x_{iD})$ is found using a current comparator. Subsequently, $i_{x_{iD}}$ and i_{x_j} are converted into pulse widths using the circuit in Fig. 6. If the buffer's threshold voltage is low, then the length of the pulse width at v_{out} measured from the rising edge of `clk` to the falling edge of v_{out} will be

$$t_w \approx \frac{i_{in}}{I_c} \frac{T_{clk}}{2}, \tag{9}$$

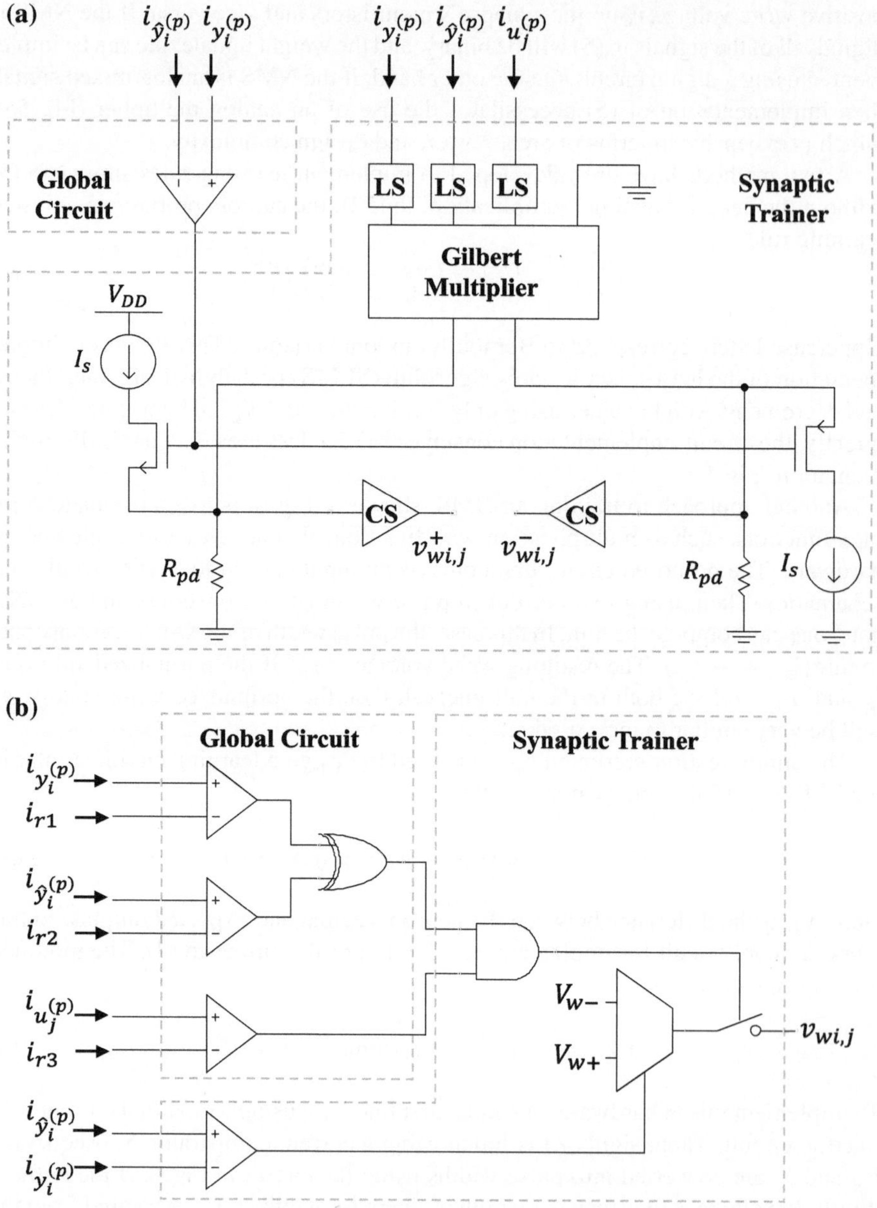

Fig. 5 Plasticity circuits for adjusting synaptic weight values in an NMS. **a** Implementation of the SLMS algorithm. **b** Implementation of the LMS algorithm

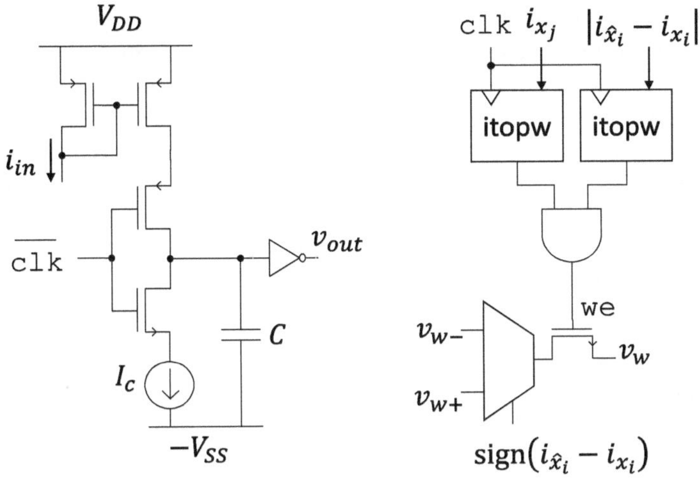

Fig. 6 Circuits for implementing the min function

where T_{clk} is the clock period. Combining two such circuits and an AND gate gives us the min function, which is used as a write enable, *WE*, signal for each synapse. Finally, the sign of the $i_{x_{iD}} = i_{\hat{x}_i} - i_{x_i}$ is used to select a positive or negative write voltage. The variation in current matching will have the largest effect on the functionality of the proposed training circuit.

The NMS training methods discussed above are just two of the many approaches that have been explored. In [39], authors explore training NMSs with the widely used backpropagation algorithm. In [54], authors present a novel training algorithm and circuit implementation for unsupervised learning in NMSs. A number of plasticity circuits for training spiking NMSs have also been designed. For example, [49] reviews several STDP implementations in NMSs. STDP is believed to be one of the primary mechanisms behind synaptic plasticity in the brain.

7 Summary and Outlook

This chapter reviewed a number of theoretical and practical considerations for the design of NMS circuits. Currently, the NMS research domain is flooded with experts in devices, circuit design, and computer architecture. More recently, groups from the neuroscience community are also becoming engaged. With so many different perspectives, the design space of NMSs is intimidatingly large, leaving a number of important design choices to be made at different levels of the design hierarchy. One question discussed in this chapter was whether NMSs should adopt a current or voltage-mode design methodology. There are clear advantages and disadvantages to both approaches, so the choice will likely be application dependent. Another question is how much biological realism should be integrated into the neuron design? Some groups have meticulously designed neuron, synapse, and plasticity circuits to mimic biological processes in the brain. To date, however, the most successful implementations of NMSs in terms of application-level performance (e.g., classification accuracy) have been based on the same high-level abstractions that are seen in the machine learning community. There should be some direct comparisons to uncover what degree of biomimicry yields the most efficient NMSs in terms of area, power, and learning capacity.

References

1. Mead, C.: Analog VLSI and Neural Systems. Addison-Wesley (1989)
2. Sinha, A., Kulkarni, M.S., Teuscher, C.: Evolving nanoscale associative memories with memristors. In: IEEE International Conference on Nanotechnology, pp. 861–864 (2011)
3. Hu, M., Li, H., Wu, Q., Rose, G.S., Chen, Y.: Memristor crossbar based hardware realization of BSB recall function. In: International Joint Conference on Neural Networks, IJCNN'12, June 2012, pp. 1–7. http://ieeexplore.ieee.org/lpdocs/epic03/wrapper.htm?arnumber=6252563
4. Kulkarni, M.S.: Memristor-based reservoir computing. In: Nanoarch, pp. 226–232 (2012)
5. Laiho, M., Lehtonen, E.: Cellular nanoscale network cell with memristors for local implication logic and synapses. In: International Symposium on Circuits and Systems, pp. 2051–2054, May 2010
6. Ebong, I.E., Mazumder, P.: CMOS and memristor-based neural network design for position detection. Proc. IEEE **100**(6), 2050–2060 (2012)
7. Adhikari, S.P., Yang, C., Kim, H., Chua, L.O.: Memristor bridge synapse-based neural network and its learning. IEEE Trans. Neural Netw. Learn. Syst. **23**(9), 1426–1435 (2012)
8. Querlioz, D., Bichler, O., Gamrat, C.: Simulation of a memristor-based spiking neural network immune to device variations. In: The 2011 International Joint Conference on Neural Networks, pp. 1775–1781, Jul 2011. http://ieeexplore.ieee.org/lpdocs/epic03/wrapper.htm?arnumber=6033439
9. Suri, M., Querlioz, D., Bichler, O., Palma, G., Vianello, E., Vuillaume, D., Gamrat, C., Desalvo, B.: Bio-inspired stochastic computing using binary CBRAM synapses. IEEE Trans. Electron Devices **60**(7), 2402–2409 (2013)
10. Merkel, C., Kudithipudi, D.: Neuromemristive extreme learning machines for pattern classification. In: International Symposium on VLSI, pp. 77–82 (2014)
11. ITRS: International technology roadmap for semiconductors (2013). http://www.itrs.net
12. Chua, L.: Memristor—The missing circuit element. IEEE Trans. Circuit Theory **CT-18**(5), 507–519 (1971)
13. Chua, L., Kang, S.-M.: Memristive devices and systems, vol. 64, no. 2 (1976)
14. Chua, L.: Resistance switching memories are memristors. Appl. Phys. A **102**(4), 765–783 (2011)
15. Yang, J.J., Strukov, D.B., Stewart, D.R.: Memristive devices for computing. Nat. Nanotechnol. **8**(1), 13–24 (2013). http://www.ncbi.nlm.nih.gov/pubmed/23269430
16. Kuzum, D., Yu, S., Wong, H.-S.P.: Synaptic electronics: materials, devices and applications. Nanotechnology, **24**(38), 382001 (2013). http://www.ncbi.nlm.nih.gov/pubmed/23999572
17. Ishigaki, T., Kawahara, T., Takemura, R., Ono, K., Ito, K., Matsuoka, H., Ohno, H.: A multi-level-cell spin-transfer torque memory with series-stacked magnetotunnel junctions. In: Symposium on VLSI Technology, pp. 47–48 (2010)

18. Waser, R., Dittmann, R., Staikov, R., Szot, K.: Redox-based resistive switching memories—nanoionic mechanisms, prospects, and challenges. In: Advanced Materials, vol. 21, no. 25–26, pp. 2632–2663, July 2009. http://doi.wiley.com/10.1002/adma.200900375
19. Ha, S.D., Ramanathan, S.: Adaptive oxide electronics: a review. J. Appl. Phys. **110**(7), 071 101–1 (2011). http://link.aip.org/link/JAPIAU/v110/i7/p071101/s1&Agg=doi
20. Yu, S., Lee, B., Wong, H.S.P.: Metal oxide resistive switching memory. In: Wu, J., Cao, J., Han, W.-Q., Janotti, H.A., Kim, H.-C. (eds.) Functional Metal Oxide Nanostructures. Springer Series in Materials Science, vol. 149, pp. 303–335. Springer, New York (2012)
21. Yang, Y., Lu, W.: Nanoscale resistive switching devices: mechanisms and modeling. Nanoscale **5**(21), 10 076–92 (2013)
22. Strukov, D.B., Williams, R.S.: Exponential ionic drift: fast switching and low volatility of thin-film memristors. Appl. Phys. A **94**(3), 515–519 (2008)
23. Yang, J.J., Pickett, M.D., Li, X., Ohlberg, D., Stewart, D.R., Williams, R.S.: Memristive switching mechanism for metal/oxide/metal nanodevices. Nat. Nanotechnol. **3**(7), 429–433 (2008)
24. Strukov, D.B., Borghetti, J.L., Williams, R.S.: Coupled ionic and electronic transport model of thin-film semiconductor memristive behavior. Small **5**(9), 1058–1063 (2009)
25. Mcdonald, N.R.: Al/CuxO/Cu memristive devices: fabrication, characterization, and modeling. Master's Thesis, SUNY Albany (2012)
26. Biolek, Z., Biolek, D., Biolková, V.: SPICE model of memristor with nonlinear dopant drift. Radioengineering **18**(2), 210–214 (2009)
27. Rak, A., Cserey, G.: Macromodeling of the memristor in SPICE. IEEE Trans. Comput. Aided Des. Integr. Circuits Syst. **29**(4), 632–636 (2010)
28. Zhang, Y., Zhang, X., Yu, J.: Approximated SPICE model for memristor. In: 2009 International Conference on Communications, Circuits and Systems, no. 5, pp. 928–931, July 2009. http://ieeexplore.ieee.org/lpdocs/epic03/wrapper.htm?arnumber=5250371
29. Batas, D., Fiedler, H.: A memristor SPICE implementation and a new approach for magnetic flux-controlled memristor modeling. IEEE Trans. Nanotechnol. **10**(2), 250–255, Mar 2011. http://ieeexplore.ieee.org/lpdocs/epic03/wrapper.htm?arnumber=5373921
30. Yakopcic, C., Taha, T., Subramanyam, G., Pino, R.: Memristor SPICE model and crossbar simulation based on devices with nanosecond switching time. In: International Joint Conference on Neural Networks, pp. 464–470 (2013)
31. Chen, Y., Wang, X.: Compact modeling and corner analysis of spintronic memristor invited paper. In: IEEE/ACM International Symposium on Nanoscale Architectures, pp. 7–12 (2009)
32. Shin, S., Kim, K., Kang, S.-M.: Compact models for memristors based on charge-flux constitutive relationships. IEEE Trans. Comput. Aided Des. Integr. Circuits Syst. **29**(4), 590–598 (2010)
33. Sheridan, P., Kim, K.-H., Gaba, S., Chang, T., Chen, L., Lu, W.: Device and SPICE modeling of RRAM devices. Nanoscale **3**(9), 3833–3840 (2011)
34. Gilbert, B.: Translinear circuits: a proposed classification. Electron. Lett. **11**(1), 14 (1975). http://digital-library.theiet.org/content/journals/10.1049/el_19750011
35. Toumazou, C., Lidgey, F.J., Haigh, D.G. (eds.): Analogue IC Design: The Current-mode Approach. Peter Peregrinus Ltd. (1990)
36. Manem, H., Rajendran, J., Rose, G.S.: Stochastic gradient descent inspired training technique for a CMOS/Nano memristive trainable threshold gate array. IEEE Trans. Circuits Syst. **59**(5), 1051–1060 (2012)
37. Soltiz, M., Member, S., Kudithipudi, D., Merkel, C., Rose, G.S., Pino, R.E.: Memristor-based neural logic blocks for non-linearly separable functions. IEEE Trans. Comput. **62**(8), 1597–1606 (2013)
38. Kim, H., Sah, M.P., Yang, C., Roska, T., Chua, L.O.: Memristor bridge synapses. Proc. IEEE **100**(6), 2061–2070 (2012)
39. Hasan, R., Taha, T.M.: Enabling back propagation training of memristor crossbar neuromorphic processors. In: 2014 International Joint Conference on Neural Networks (IJCNN), pp. 21–28. IEEE, July 2014. http://ieeexplore.ieee.org/lpdocs/epic03/wrapper.htm?arnumber=6889893

40. Indiveri, G., Linares-Barranco, B., Hamilton, T.J., van Schaik, A., Etienne-Cummings, R., Delbruck, T., Liu, S.-C., Dudek, P., äfliger, P. H., Renaud, S., Schemmel, J., Cauwenberghs, G., Arthur, J., Hynna, K., Folowosele, F., Saighi, S., Serrano-Gotarredona, T., Wijekoon, J., Wang, Y., Boahen, K.: Neuromorphic silicon neuron circuits. Front. Neurosci. **5**, 73 (2011). http://www.pubmedcentral.nih.gov/articlerender.fcgi?artid=3130465&tool=pmcentrez&rendertype=abstract
41. Rosenblatt, F.: The perceptron: a probabilistic model for information storage and organization in the brain. Psychol. Rev. **65**(6), 386–408 (1958). http://www.ncbi.nlm.nih.gov/pubmed/13602029
42. Minsky, M., Papert, S.: Perceptrons—Expanded edition: An Introduction to Computational Geometry. MIT Press (1987)
43. Kandel, E.J., Schwartz, J. H., Jessell, T.J., Siegelbaum, S.A., Hudspeth, A. J.: Principles of Neural Science, 5th edn. McGraw Hill (2013)
44. Soudry, D., Castro, D. D., Gal, A., Kolodny, A., Kvatinsky, S.: Memristor-based multilayer neural networks with online gradient descent training. IEEE Trans. Neural Netw. Learn. Syst. 1–14 (2015)
45. Karlik, B.: Performance analysis of various activation functions in generalized MLP architectures of neural networks. Int. J. Artif. Intell. Expert Syst. **1**(4), 111–122 (2015)
46. Wilamowski, B.M.: Understanding neural networks. In: Wilamowski, B.M., Irwin, J.D. (eds.) Intelligent Systems (Industrial Electronics), 2nd edn., Chap. 5, pp. 5–1. CRC Press (2011)
47. Merkel, C., Kudithipudi, D., Sereni, N.: Periodic activation functions in memristor-based analog neural networks. In: International Joint Conference on Neural Networks, pp. 1–7 (2013)
48. Attwell, D., Laughlin, B.: Energy budget for signaling in the grey matter of the brain. J. Cereb. Blood Flow Metab. Off. Int. Soc. Cereb. Blood Flow Metab. **21**(10), 1133–1145 (2001). http://www.ncbi.nlm.nih.gov/pubmed/11598490
49. Serrano-Gotarredona, T., Masquelier, T., Prodromakis, T., Indiveri, G., Linares-Barranco, B.: STDP and STDP variations with memristors for spiking neuromorphic learning systems. Front. Neurosci. **7**, 2 (2013). http://www.pubmedcentral.nih.gov/articlerender.fcgi?artid=3575074&tool=pmcentrez&rendertype=abstract
50. Panzeri, S., Brunel, N., Logothetis, N.K., Kayser, C.: Sensory neural codes using multiplexed temporal scales. Trends Neurosci. **33**(3), 111–120 (2010). http://www.ncbi.nlm.nih.gov/pubmed/20045201
51. Purves, D., Augustine, G.J., Fitzpatrick, D., Hall, W.C., LaMantia, A.-S., White, L.E.: Neuroscience, 5th edn. Sinauer Associates, Inc. (2012)
52. Merkel, C., Kudithipudi, D.: A stochastic learning algorithm for neuromemristive systems. In: System on Chip Conference, pp. 359–364 (2014)
53. Widrow, B.: An adaptive "ADALINE" neuron using chemical "Memistors". Stanford University, Tech. Rep. (1960)
54. Merkel, C., Kudithipudi, D.: IEEE National Aerospace and Electronics Conference, NAECON'15. IEEE (2015)

Memristor-Based Platforms: A Comparison Between Continous-Time and Discrete-Time Cellular Neural Networks

Young-Su Kim, Sang-Hak Shin, Jacopo Secco, Keyong-Sik Min and Fernando Corinto

1 Introduction

In this chapter, theory, circuit design methodologies and possible applications of Cellular Nanoscale Networks (CNNs) exploiting memristor technology are reviewed. Memristor-based CNNs platforms (MCNNs) make use of memristors to realize analog multiplication circuits that are essential to perform CNN calculation with low power and small area. Compared to memristor-based crossbar architectures proposed to mimic fundamental neuron–cell-level operations (e.g. Spike Time Dependent Plasticity), MCNNs are more suitable in mimicking the mammalian visual system that processes topographic image flows through a set of separate spatial–temporal channels. As so, it will also be shown how special classes of MCNNs, for instance Continous-Time and Disctrete-Time CNNs (DTCNNs) can be assimilated to other algorithmic techniques for image and data processing such as Cellular Automata (CA), with fulfilling applicative examples. In the chapter, it will be presented a detailed comparison of the two circuitry designs and their algorithmic interactions that occur to obtain similar results on generic digital images. The scope is to understand the importance of the use of MCNNs and the actual technical benefits that derive from these systems. The chapter summarizes the basic concepts of CNN computation at first, and memristor-based CNN circuits that are very useful in performing analog multiplication. Some practical issues of MCNN circuits and their possible solutions are outlined as well. Finally, we present the results of MCNNs calculation with Laplacian template that is used for edge detection in various image processing tasks.

Y.-S. Kim · S.-H. Shin · J. Secco · K.-S. Min · F. Corinto (✉)
Politecnico di Torino, Corso Duca degli Abruzzi 24, 10129 Torino, Italy
e-mail: fernando.corinto@polito.it

J. Secco
e-mail: jacopo.secco@polito.it

© Springer (India) Pvt. Ltd. 2017

M. Suri (ed.), *Advances in Neuromorphic Hardware Exploiting Emerging Nanoscale Devices*, Cognitive Systems Monographs 31,
DOI 10.1007/978-81-322-3703-7_4

2 Backgorund

One of the neuromorphic applications that use memristors is CNNs [1–4]. In the memristor-based CNNs, synaptic weights are calculated by analog multiplication which can consume smaller power and occupy smaller area than digital multiplication [5–8]. Figure 1 shows a conceptual diagram of analog multiplication of input matrix and template matrix. In the matrix multiplication, the input matrix, $[X]$ is multiplied by the CNN template, $[A]$, as shown in Fig. 1. By multiplying $[X]$ with $[A]$, we can calculate the output matrix, $[Z]$. Here, we assume that the input matrix has 60×60 pixels, and the CNN template is composed of nine coefficients from a_0 to a_8. As shown in Fig. 1, one 3×3 sub-matrix in $[X]$ that is represented by X_1 is multiplied with the CNN template, $[A]$. Thereby, we can obtain one output pixel, Z_1 in the output matrix, $[Z]$. Similarly, X_2 is multiplied with A, and we can calculate Z_2. And, also, X_3 is multiplied with A. Z_3 can be given as an output of this sub-matrix multiplication of X_3 with A.

The analog multiplication of sub-matrix and CNN template is based on Ohms law. From Ohms law, the memristors voltage is given by $v(t) = M(t)i(t)$, where $v(t)$ and $i(t)$ are the memristors voltage and current, respectively. $M(t)$ is the memristance that can be varied dynamically with respect to time, according to the history of applied current and voltage [8]. Here, if the input matrix is the input current applied to memristor, we can regard memristors voltage, $v(t)$ as the multiplication result of

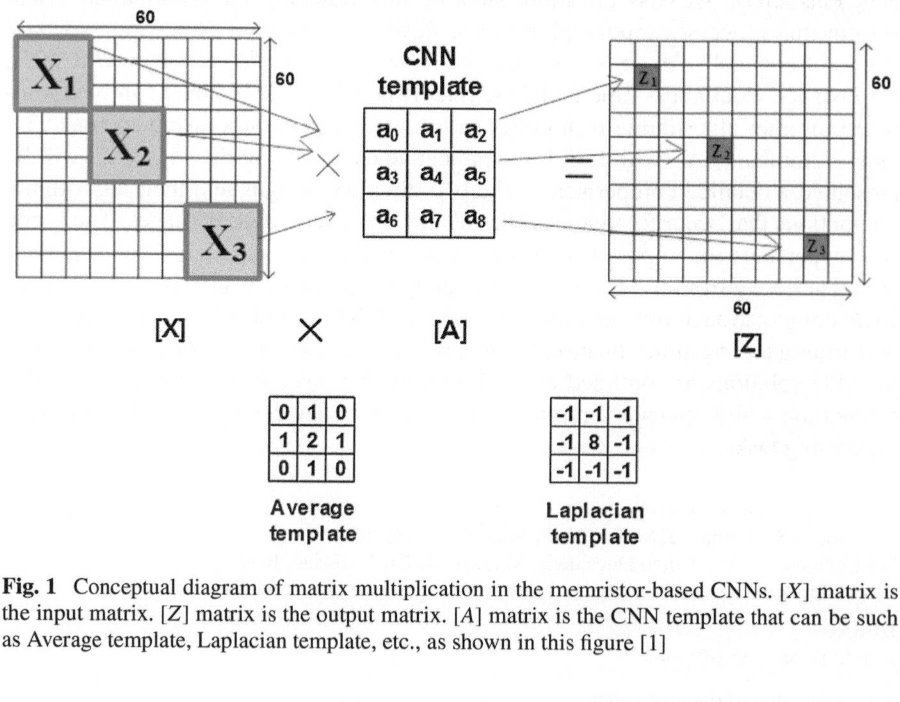

Fig. 1 Conceptual diagram of matrix multiplication in the memristor-based CNNs. $[X]$ matrix is the input matrix. $[Z]$ matrix is the output matrix. $[A]$ matrix is the CNN template that can be such as Average template, Laplacian template, etc., as shown in this figure [1]

$M(T)$ and $i(t)$. Here, the CNN template is composed of nine different memristance values of $M(T)$ from nine memristors.

Unfortunately, the fact that memristance value of $M(t)$ is changed dynamically causes an important problem in matrix multiplication. As we repeat the multiplication over and over, memristance values of $M(t)$ can be drifted away from the originally programmed values. It means that the coefficients of CNN template become different from the original values. Though an amount of memristance drift is negligible for one-time multiplication, memristance may be varied very much as matrix multiplication is repeated over and over. Hence, a kind of memristance restoring circuit is indispensable in sustaining the multiplication accuracy during the CNN operation. In this work, we propose a new memristance restoring circuit to recover the drifted memristance values to the original ones.

3 New Memristance Restoring Circuit

Figure fk2a shows the synaptic weighting circuit of one CNN cell that has one current mirror, one common-mode feedback circuit, nine differential pairs for nine input currents and nine memristors [1]. The current mirror is composed of MP_3 and MP_4 and R_L means the load resistor. The common-mode feedback biasing circuit adjusts the tail-current biasing voltage for nine differential pairs to keep V_{OUT} around $V_{DD}/2$ when the input currents are zero, regardless of Process–VDD–Temperature variations. The differential pairs in Fig. 2a convert the differential voltages that are obtained by memristor weighting circuits to the output currents for the summation of nine output currents. The nine memristor weighting circuits can calculate nine products of the input currents and the coefficients of CNN template.

In Fig. 2a, M_{W0} is the weighting memristor for calculating the multiplication of the input current I_{IN0} with the a_0 coefficient of CNN template. S_0 and S_{B0} are the complement switches. When S_0 is turned on, S_{B0} is turned off and vice versa. When S_0 is on, I_{IN0} is applied to M_{W0} to calculate the multiplication of I_{IN0} with M_{W0}. If the coefficient value in CNN template matrix is zero such as a_0 of CNN average template, S_{B0} becomes on and S_0 becomes off. This is because we cannot make memristance value zero. Hence, instead of making the memristance value zero, we can turn off S_0 and turn on S_{B0} to make zero the output current of the multiplication of I_{IN0} with a_0. One more thing to note here is that the weighting memristor, M_{W0}, can be connected to M_{N4} or M_{N5} according to the polarity of coefficient. If the coefficient in CNN template is positive, M_{W0} is connected to M_{N4}. If the coefficient is negative, M_{W0} should be connected to M_{N5}. Here, we assume the Laplacian template, where a_0 is negative. Thus, M_{W0} is connected to M_{N5} in Fig. 2a. On the contrary, M_{W4} is connected to M_{N7} not M_{N8}, because a_4 in CNN Laplacian template is positive. The nine differential pairs convert nine results of matrix multiplication into the currents. These currents are summated by the current mirror that is composed of M_{P3} and M_{P4}. The summated current is delivered to RL to make the final output voltage. M_{P1}, M_{P2},

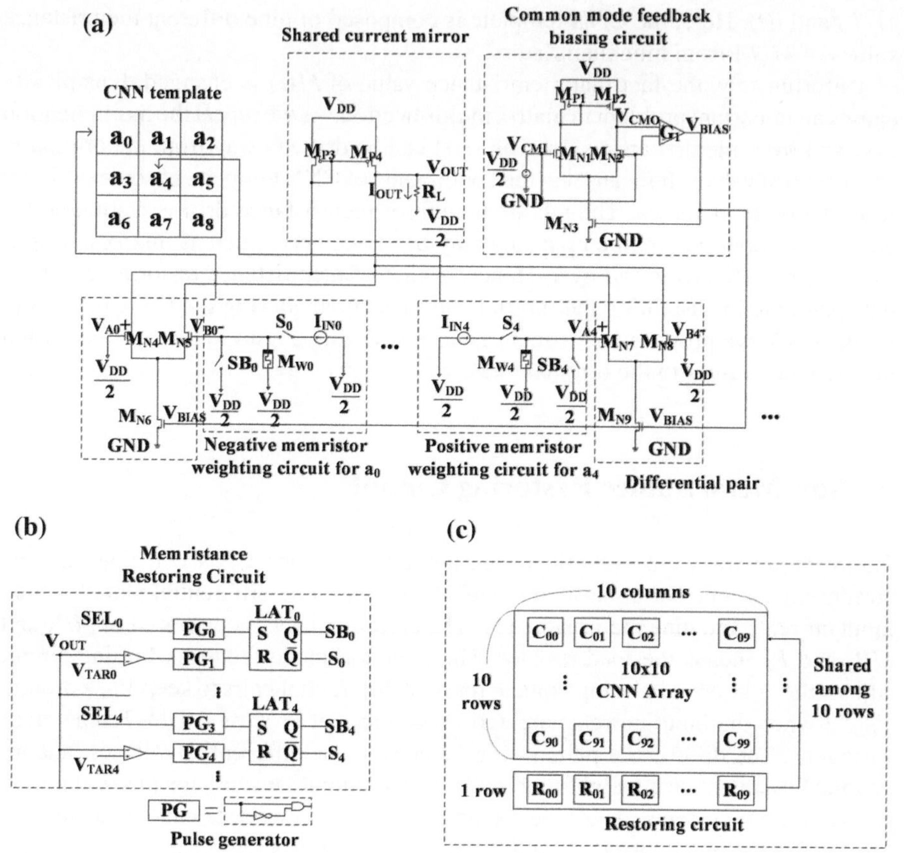

Fig. 2 **a** The synaptic weighting circuit with memristors for one CNN cell. **b** The memristance restoring circuit for one CNN cell. **c** The one row of restoring circuits can be shared among 10 rows of synaptic weighting circuits in 10 × 10 CNN cell array [1]

M_{N1} and M_{N2} constitute the common-mode feedback biasing circuit to fix the output voltage by $V_{DD}/2$ when the input currents are zero.

The memristance restoring circuit for one CNN cell is shown in Fig. 2b. In one CNN cell, there are nine memristors to be restored. Thus, we have nine memristance restoring circuits. SEL_0 means the selection signal for the 0th memristor, M_{W0} in Figure 2a. SEL_4 means a signal to select the 4th memristor, M_{W4}. Here, V_{OUT} is the output voltage of the synaptic weight circuit in Fig. 2a. V_{TAR0} is a target voltage for restoring the 0th memristor. V_{TAR4} is a target voltage for the 4th memristor. In Fig. 2b, *PG* is the pulse generator circuit, and *SR* latch is used to control the switches in the memristor weighting circuit. Here '$/Q$' and 'Q' are the outputs of *SR* latch, LAT_0. They are used to control S_0 and SB_0 in the memristor weighting circuit of a_0. Similarly, the outputs of *SR* latch, LAT_4 control S_4 and SB_4 in the memristor circuit of a_4. For restoring MW_0 to the original programmed value, 'RESET' pulse is generated in PG_1, if V_{OUT} becomes equal to V_{TAR0}. By doing so, the generated pulse can go into the *SR* latch, LAT_0, to stop programing M_{W0} further beyond the original programmed value.

Figure 2c shows the block diagram of CNN circuits with 10 × 10 cells and one row of memristance restoring circuits. $C_{0,0}$ is the CNN cell of $ROW = 0$ and $COL = 0$. $C_{0,9}$ is the CNN cell of $ROW = 0$ and $COL = 9$. Similarly, $C_{9,0}$ and $C_{9,9}$ are the cell of $ROW = 9$ and $COL = 0$ and the cell of $ROW = 9$ and $COL = 9$, respectively. The memristance restoring circuits are shown below in Fig. 2c. The 0th restoring circuit, R_0 is for the 0th column of CNN array. The 9th restoring circuit, R_9 is for the 9th column. Each restoring circuit can restore one CNN cell.

The more detailed schematic of Fig. 2c is shown in Fig. 3. Here, we assume a 10 × 10 CNN array. $CELL_{0,0}$ is the CNN cell of $ROW = 0$ and $COL = 0$. $CELL_{0,9}$ is the CNN cell of $ROW = 0$ and $COL = 9$. Similarly, $CELL_{9,0}$ is the cell of $ROW = 9$ and $COL = 0$. $CELL_{9,9}$ is the cell of $ROW = 9$ and $COL = 9$. Each CNN cell is composed of nine weighting memristors, nine differential pairs, one current mirror, etc.,

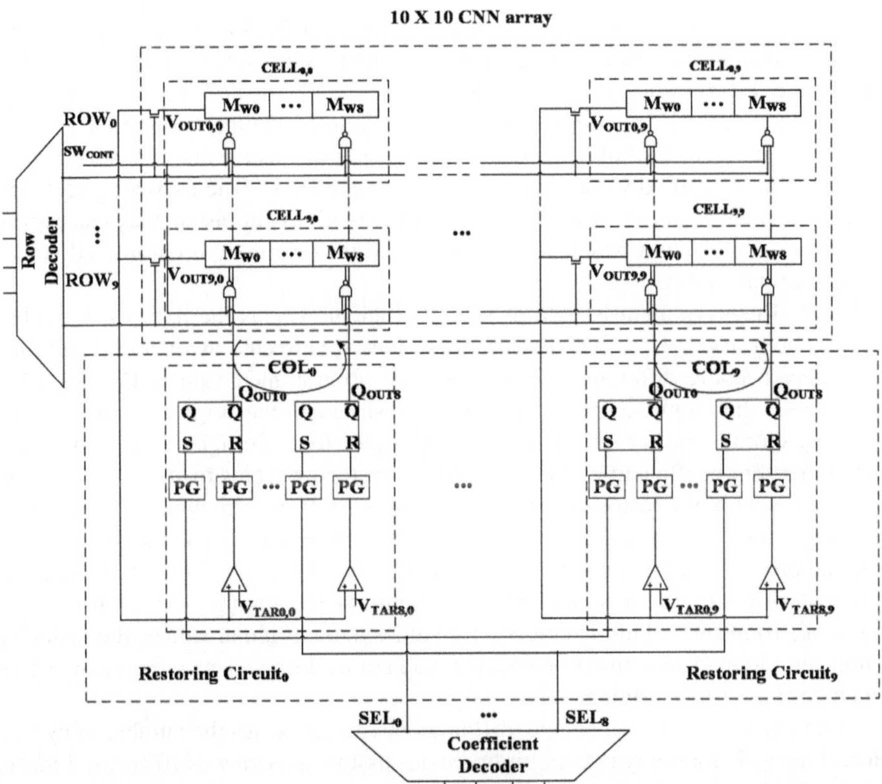

Fig. 3 10 × 10 Synaptic weighting circuits and 10 memristance restoring circuits that are shared among 10 rows in 10 × 10 CNN cell array

as shown in Fig. 2a. M_{W0} and M_{W8} are representing the first coefficient and the ninth one of the CNN template, respectively. $V_{OUT0,0}$ and $V_{OUT0,9}$ are the output voltages of $CELL_{0,0}$ and $CELL_{0,9}$, respectively, for $ROW = 0$. $V_{OUT9,0}$ and $V_{OUT9,9}$ are the output voltages of $CELL_{9,0}$ and $CELL_{9,9}$, respectively, for $ROW = 9$. At the bottom row in Fig. 3, the memristance restoring circuit of $COL = 0$ is connected to the output voltages of $COL = 0$. Similarly, the memristance restoring circuit of $COL = 9$ can recover the memristance values which are in $COL = 9$ to the originally programmed values. As shown in Fig. 3, the memristance restoring circuits are shared among ten rows from $ROW = 0$ and $ROW = 9$. This sharing of memristance restoring circuits is very helpful in reducing the layout area of memristance restoring circuits.

4 Simulation Results

At this section, we discuss the simulation results of the memristance restoring circuit. The circuit simulation is done by CADENCE SPECTRE [9] using HP memristor model [4, 8] and SAMSUNG CMOS $0.13m$ SPICE model. Figure 4a shows that the random input currents are applied to weighting memristors for the first 0–100 cycles. During the 0–100 cycles, the memristance values can be drifted from the originally programmed values. And, during the 100–150 cycles, NO Operation (NOP) is executed in the CNN array. For the 100–150 cycles, memristance values are not changed further from the drifted values during the 0–100 cycles. For the following 150–350 cycles, the memristance restoring circuits start to work. By doing so, they can restore the memristance values which were drifted from the originally programmed values to the original values.

Figure 4b shows memristance of four coefficients versus the number of cycles. During the 0–100 cycles, the memristance values are affected by the input currents; thus, they become different from the original programmed values. For the 150–350 cycles, the memristance values can be restored to the original values by the restoring circuits. In Fig. 4b, the restoring times of four coefficients seem different each other. For coefficient 2, the memristance restoring seems to start at 175 cycle. For coefficient 4, the restoring circuit seems to work from 225 cycle. This different restoring time is because the restoring circuit can restore only one memristor to its original programmed value among nine memristors of nine coefficients at one time. The restoring circuit can sense only one memristance among nine memristors at one time. To restore all nine memristance values to the original values, the restoring circuit should work nine times one by one. Each time, the restoring circuit can restore one memristor to the original value.

Figure 4c shows the percentage drift in memristance versus the number of cycles. Here, one cycle for doing one multiplication is assumed as long as 100 ns. In Fig. 4c, during the 0–100 cycles, we can know that the memristance values are drifted from the originally programmed values by the random input currents, as mentioned in Fig. 4b. One more thing to note in Fig. 4c is that the percentage drift in memristance is depending on its coefficient value and the amount of input current applied to this

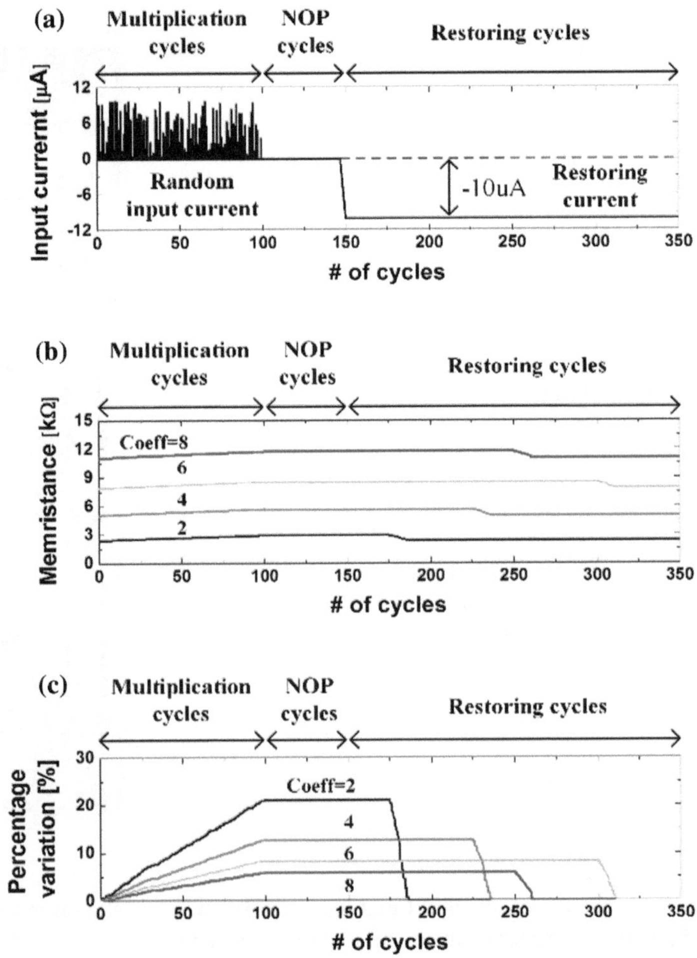

Fig. 4 **a** The random input currents are applied during the 0–100 cycles. For the 100–150 cycles, No operation is performed in CNN array. For the following 150–350 cycles, the memristance restoring circuits work to restore the memristance values to the original programmed values. **b** Memristance variation in the synaptic weighting circuit with increasing the number of cycles. **c** Percentage variation in memristance in the synaptic weighting circuit with increasing the number of cycles [1]

memristor. In Fig. 4c, coefficient 2 shows the largest amount of percentage drift in memristance among four coefficients. If the coefficient value becomes larger, the percentage drift in memristance becomes smaller.

Figure 5a shows the edge-detected Lena image with 60 × 60 pixels which is calculated by CNN Laplacian template. Here, we can see the detected edges are shown in white, while the rest parts of Lena image are in black. Here, if we assume that memristors in CNN array are applied by the random input current as long as 100

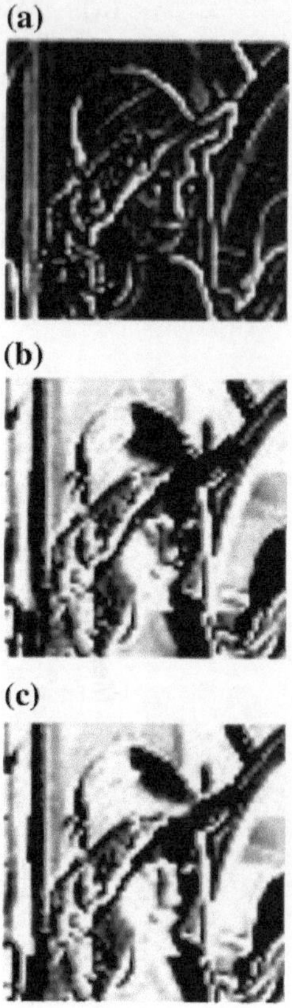

Fig. 5 **a** The edge-detected Lena image with 6060 pixels that is calculated by CNN Laplacian template. **b** The image degradation of the edge-detected Lena image due to memristance drift with 10 % variation. **c** The image degradation of the edge-detected Lena image due to memristance drift with 20 % variation. Here, the memristance drift is caused by the random input currents which are applied to memristors in CNN array as long as 100 cycles

operating cycles, the edge-detected Lena image is degraded as shown in Fig. 5b and c. In Fig. 5b and c, the variation in the drifted memristance is assumed 10 % and 20 %, respectively. Comparing Fig. 5b and c, the degradation of the detected edges in Lena image seems more severe in Fig. 5c than Fig. 5b. Using the memristance restoring circuit in Fig. 3, we can restore the drifted memristance to the original programmed values to avoid the image degradation that is shown in Fig. 5b and c. Finally, it should be noted that the memristance restoring circuit can be applied to not only CNN template, but also the other CNN templates such the average template, etc.

5 Cellular Automata and DTCNNs

A cellular automaton (CA) is a discrete time-space model used to study computability theory [10]. Furthermore, as described in details in [11], memristive CNNs, in particular DTCNNs, can be assimilated to memristive CAs. It was first theorized by Satnslaw Ulham and John von Neumann in the late 1940s and later defined by

Stephen Wolfram. This assimilation is possible since memrisitive CAs are able to perform similar functions as neural networks due to their dynamic nonlinear structure defined by coupling identical simple dynamical elements called cells. Since 2008, when R. Stanley Williams from the Hewlett-Packard Company was able to fabricate a nanoscale memristor [4], a nonlinear resistive device capable of retaining memory of its internal state [12], the scientific impact of such technologies has grown.

Cellular automata consist in a regular grid of cells, each one with a finite number of states. The grid must be of finite dimensions. The states of the cells composing the grid at time instant t = 0 are the initial states of the automaton and are arbitrarily set. At each time point, the combination of the cell's states forms a generation. The generational evolution occurs at specific discrete time instants, and it depends on the state of the previous generation. More precisely, each cell changes its state from a generation to another depending on its previous state and/or on the states of the surrounding cells. The possible number of rules that the system can implement strongly depends on surrounding radius that effects the change of the given cell. Supposing to have an array of n cells and the generational evolution of a given cell of the system is given by its actual state (1 or 0) and the state of the two adherent cells then there are $2^3 = 8$ possible combinations that lead to 2^8 possible rules.

From Wolfram's definition, a system must present the following characteristics in order to be defined a CA [10]:

- there must be a spatial representation of the involved entities;
- uniformity: or in other words all the entities must have the same characteristics and must be identical other than interchangeable; and
- locality: each entity changes its state from a generation to the other taking into account the states of the entities within a given surrounding radius.

Still, from Wolfram's theory, CAs can be described as a fourfold $< d, Q, N_n, f >$ where

- d is the dimension of the CA;
- Q is the space of the states which the cells can assume;
- N_n is the neighbourhood index which describes the region of influence of the other cells for the given cell's state change; and
- f is the generation transition function which describes the state change of each cell at each time instant $t = \tau + nT$ and must be a function of a cell neighbourhood described by parameter N_n.

Itoh and Chua [11] first described an implementation of a cellular automaton with inputs using memristors. The dynamics of their system were given by

$$\begin{aligned} y_{i,j}(nT) = M(&\sum_{g,l\in(-1,0,1)} a_{g,l}y_{i+g,j+l}((n-1)T)+ \\ &+ \sum_{g,l\in(-1,0,1)} b_{g,l}u_{i+g,j+l}((n-1)T) + \Delta T) \end{aligned} \tag{1}$$

where y is the output or in other words the state of the cell, and u are the external inputs of the system. T is the time period in which there is a new input and, therefore, a new generation change in the system, while Δt represents the charge accumulation in the memristor during the previous generations. $a_{g,l}$ and $b_{g,l}$ are elements of two distinct matrices A and B which both have size $G \times L$. A is the template that contains the weights given to the neighbour cell states (feed-forward), and B contains the weights given to the external input (feed-back). A and B are the necessary sufficient elements, which describe the imposed rules to the memristor CA. Function $M(\cdot)$ denotes the memristace change function.

6 Belief Propagation Inspired Algorithm and Cellular Automaton Equivalence for RGB Image Processing

In the previous section, an algorithm was introduced by which Itoh and Chua managed to control the memristances of each cell [11]. The CA that is presented hereafter exploits the BPI algorithm, and the cells have been assimilated with mathematical models of memristors (see [13, 14] for further details). The BPI algorithm [15] works throughout a single layer of cells called synapses. Each synapse returns a binary value, which depends on the actual memristance values of each cell. For the BPI, it is necessary to compute the "normalized" memristance of each cell $h_i \in [-1, 1]$ subdivided in k discrete steps. The highest possible resistance of the single element (R_{off}) corresponds to $h_i = -1$, and viceversa the lowest possible value (R_{on}) corresponds to $h_i = 1$. All the normalized memristances are to be commuted into synaptic weights as $w_i = \frac{1}{2}(sign(h_i) + 1)$, where $sign(\cdot)$ represents a signum function. According to the given rule intended to be learned by the system, it is necessary to set a threshold parameter $\theta \in [0, N]$, where N is the number of cells composing the array. Given a set of binary input patterns ζ_i, and for each pattern a desired output σ_d (also binary) that follows the implemented rule, it is possible to calculate the total current flowing from the cells as $I = \sum_{i=1}^{N} w_i\zeta_i$. Once computed the "stability parameter" as $\Delta = (2\sigma_d - 1)(I - \theta)$, the evolutionary change may be described by the three following rules:

(R1) if $\Delta \geq 0$, then all $w_i^\tau = w_i^{(\tau+1)}$;
(R2) if $\Delta < 0$, then all $h_i^{\tau+1} = h_i^\tau + 2\zeta_i^\tau(2\sigma^\tau - 1)$; and
(R3) with probability $p_s \in [0, 1]$, if $w_i^\tau > 0$ then $h_i^{\tau+1} = h_i^\tau - 2\zeta_i^\tau$; else $h_i^{\tau+1} = h_i^\tau + 2\zeta_i^\tau$.

In order to find an equivalence between the CA described in Eq. 1 and a CA built using the BPI, several arrangements were made on the parameters of the algorithm. The probability p_s was set to zero in order to have total control of the generation changes of the states of the cells composing the system. The number of states k was set equal to 2 in order to have $h_i \in \{-1, 1\}$ significantly cutting down the computational complexity. Combining $R1$ and $R2$ with the definition of stability parameter Δ function $J(\cdot)$ can be obtained, which describes the evolution dynamics of the cells:

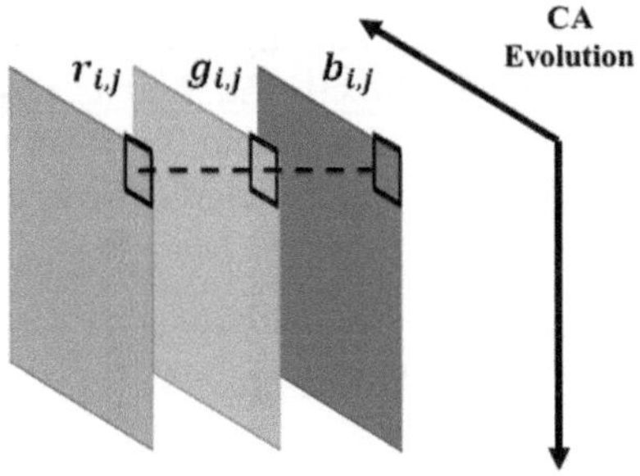

Fig. 6 Brief depiction of the crosswise CA. All three matrices (R, G and B) are divided in cells which are compared through the equations in $G(r_{i,j}, g_{i,j}, b_{i,j})$. The results of the system give the inputs to the CA, which evolves horizontally and vertically from the original neighbourhood array until it covers the whole image

$$J_i(\zeta, \sigma_d) = h^{\tau+1} = h_i^\tau + 2\zeta_i^\tau \frac{(sign(\Delta) - 1)}{2sign(I - \theta)} \tag{2}$$

τ in $R1$, $R2$, $R3$ and $J(\cdot)$ denotes the time point of a given generational change. As it is possible to see from Eq. 2, $J(\cdot)$ is the function that states the weights given to the inputs (ζ_i and σ_d).

In our BPI-CA, the actual cells that evolve during the process are posed in three memristor-composed matrices with the same size of the analyzed image (H_r, H_g and H_b). This is so since differently from normal CA use (grey tone image processing and filtering), the BPI-CA can be used to elaborate full colour map images (RGB). As an initial condition, the memristive elements were all set to the R_{off} state ($h_{H_r,H_g,H_b|i,j} = -1$). The iterations of the system do not occur in the single matrix but crosswise in the three memristive blocks, taking into account the spacial correspondence of the cells as shown in Fig. 6. The neighbourhood in which the region of influence is described is composed by the triplets of cells that correspond to the same position in the three memristive matrices ($N_n = 3$). Considering Eq. 1, it is possible to rewrite Eq. 2 in the same form.

$$\begin{aligned} h_{x|i,j}^{\tau+1} &= M(\sum_{x=1}^{N_n} a_{x|i,j} h_{x|i,j}^\tau + \sum_{x=1}^{N_n} \zeta_{x|i,j}^\tau (2\frac{(sign(\Delta) - 1)}{2sign(I - \theta)})) \\ &= M(\sum_{x=1}^{N_n} a_{x|i,j} h_{x|i,j}^\tau + \sum_{x=1}^{N_n} J_{x|i,j}(\zeta, \sigma_d) + \Delta t) \end{aligned} \tag{3}$$

In Eq. 3,[1] the threshold parameter was set to $\theta = 2.5$. All the generation changes in the system must be input-based, so in the case of our CA performing the BPI, the

[1] In Eq. 3 the cell states are noted as $h_{i,j}$ instead if $y_{i,j}$ since for the BPI algorithm the possible states are discrete and properly fixed. On the other hand, in the algorithm by Itoh and Chua the state of the single cell coincides with the actual internal state of the memristive element.

matrix that gives the weights to the surrounding states is $A = [0 \quad 1 \quad 0]$. By inserting A in Eq. 3, it is possible to obtain directly Eq. 2. It is possible to notice that the evolutionary algorithms described by Eqs. 1 and 3 present the same properties of a CA as described by Wolfram, thus are considered to be equivalent.

7 Element Detection in RGB Image

The BPI-CA can be used as aforementioned in the processing of generic RGB images. The scope for which ha been performed in this work is the detection of particular elements from digital images. In order to obtain a good identification of a specific element depicted in a generic coloured image, it is necessary to understand the properties of the image segment that shows the element itself. It is necessary to understand that the BPI-CA was performed on pictures taken with generic digital optic devices such as the one shown in Fig. 7a. Every element of a picture has a specific colour combinations that diverges from the rest of the background. Considering a generic RGB image, it is possible to decompose its three colour maps. Considering the single elements of R, G and B, it is possible to find mathematical relations between the three, here described as $G(r_{i,j}, g_{i,j}, b_{i,j})$ (as described in Fig. 6), in order to distinguish the patterns that are only proper of the areas of the depicted detail that is intended to be isolated. $G(\cdot)$ is a system of equations that take inspiration from the "green screen" detection techniques and its equations vary according to different factors relating the image, i.e. the exposure of the picture, but can be adjusted through digital image filters that act directly on the brightness histogram.

All the elements of the RGB matrices of the picture (i.e. Fig. 7a) were analyzed according to these functions and the solutions of the equations gave the corresponding value range in which all three must reside in order to identify the particular element obtaining $G(r_{i,j}, g_{i,j}, b_{i,j}) = \{\zeta_{r|i,j}, \zeta_{g|i,j}, \zeta_{b|i,j}\}$. In other words, the colours that identify the object are singular and are not repeated in other areas of the picture external to the object itself. As mentioned before, the BPI-CA does not perform its iterations singularly in each matrix, but it evolves crosswise between R, G and B as shown in Fig. 6.

Each element of the three matrices was considered as an input cell. The environmental neighbourhood of the generational evolution of the cells is given by the spatial relation of the elements of the three matrices. The BPI other than cellular external inputs needs the according desired output in order to arrange the eventual generation change of the states. By these means, the rule imposed to the CA can be easily described by Table 1.

Once all the iterations are over and all the cells in the three H matrices have been set according to the algorithm, matrix S was created re-presenting all the inputs to the cells. In this second computational phase, the cells do not change their states but retain memory of the CA interactions. S has the same size of the image and its elements are the outputs of the BPI calculated as:

Table 1 BPI–CA Particular colour detection rule

$\zeta_{r\|i,j}$	$\zeta_{g\|i,j}$	$\zeta_{b\|i,j}$	$\sigma_{d\|i,j}$
0	0	0	0
0	0	1	0
0	1	0	0
0	1	1	0
1	0	0	0
1	0	1	0
1	1	0	0
1	1	1	1

$$s_{i,j} = \Theta(w_{H_r|i,j}\zeta_{r|i,j} + w_{H_g|i,j}\zeta_{g|i,j} + w_{H_b|i,j}\zeta_{b|i,j} - \theta). \quad (4)$$

$\Theta(\cdot)$ is the Heaveside function, $w_{H|i,j}$ are the synaptic weights computed from the cells composing H_r,H_g and H_b, and θ is the threshold parameter.

The resulting contents of S matrix corresponds to a bit-wise image representing the mask of the element since the white pixels ($s_{i,j} = 1$) are spatially corresponding to the ones in the native image in which the object is present as shown in Fig. 7. It is possible to notice that in the particular case of Fig. 7, the subject has two very close objects with the same colour scheme and that the system was able to exclude the rest of the background of the picture, detecting in fact, both elements. The BPI-CA is able to precisely detect the contours of a single chosen element. The results depend on the conditions present when the picture was taken. All images used to prove the reliability of the method were randomly selected from several databases and all have variable features, i.e. use of the flash when capturing the picture and degree of focus.

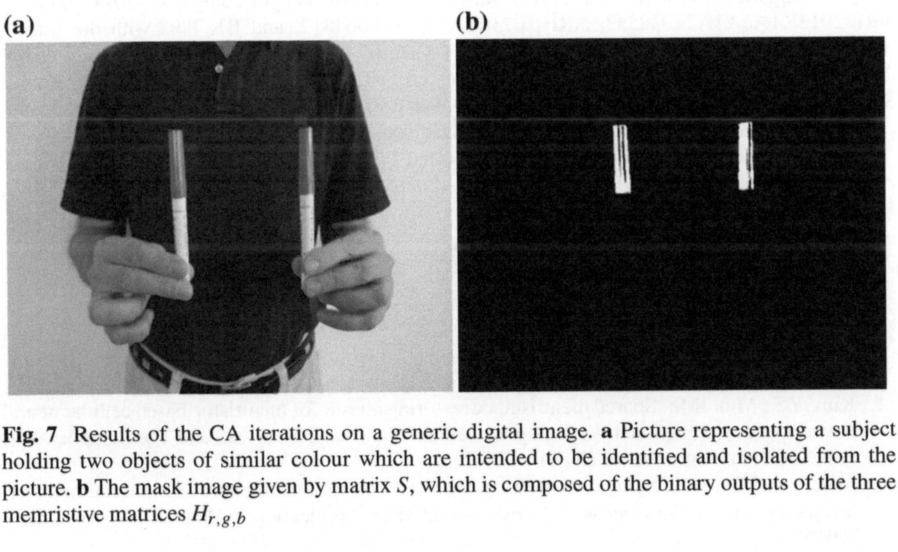

Fig. 7 Results of the CA iterations on a generic digital image. **a** Picture representing a subject holding two objects of similar colour which are intended to be identified and isolated from the picture. **b** The mask image given by matrix S, which is composed of the binary outputs of the three memristive matrices $H_{r,g,b}$

Although this variability could lead to aleatory results, applying well-known image filtering solutions (same filters for each picture examined), the system was able to detect several elements with the same high precision for all pictures.

8 Conclusions

In the first part of this chapter, the memristance restoring circuit in the memristor-based CNN array was proposed and verified. The restoring circuit can be shared among 10 different rows of synaptic circuits in CNN array to minimize the area overhead, when we assume 10×10 CNN array. If we share the restoring circuit among 10 rows in 10×10 CNN array, the area overhead of the restoring circuit can be reduced to 1/10. Moreover, Lena image with 60×60 pixels was calculated with Laplacian CNN template to detect the edges of Lena image. In the detected edges of Lena image, we could verify the image degradation due to memristance drift. Here, the memristance drift was caused by the random input currents which were applied to memristors in CNN array as long as 100 cycles.

On the other hand in the second part, a possible application for the implementation of image precessing device though CA is presented. The aim of this work is to demonstrate the possibility to exploit neuromorphic systems, such as CAs or CNNs, in various fields that require large amount of data storage and processing with low hardware use. Memristive devices could provide breakthroughs in the analysis of all sorts of data, yielding highly efficient methods to measure various parameters. The system presented herein is highly reproducible and can easily be used in conjunction with other technologies. To conclude, the future development and use of this system is likely to lead to improvements in industrial and high automation fields.

Acknowledgements The first part of this work was financially supported by NRF-2011-0030228, NRF-2013K1A3A1A25038533, NRF-2013R1A1A2A10064812, and BK Plus with the Educational Research Team for Creative Engineers on Material-Device-Circuit Co-Design (Grant No: 22A20130000042), funded by the National Research Foundation of Korea (NRF), and by Global Scholarship Program for Foreign Graduate Students at Kookmin Univ. The CAD tools were supported by IC Design Education Center (IDEC), Daejeon, Korea.

The second part of this work has been supported by the Ministry of Foreign Affairs *Con il contributo del Ministero degli Affari Esteri, Direzione Generale per la Promozione del Sistema Paese*.

References

1. Kim, Y.S., Min, K.S.: Shared memristance restoring circuit for memristor-based cellular neural networks. In: International Workshop on Cellular Nanoscale Networks and Their Applications (CNNA2014), Notre Dame, IN, July 2014
2. Kim, Y.S., Min, K.S.: Synaptic weighting circuits for cellular neural networks. In: International Workshop on Cellular Nanoscale Networks and Their Applications (CNNA2012), Turin, Italy (2012)

3. Chua, L.O., Yang, L.: Cellular neural networks: theory. IEEE Trans. Circ. Syst. **35**(10), 1257–1272 (1998)
4. Strukov, D.B., Snider, G.S., Stewart, D.R.: The missing memristor found. Stanley R. Nat. **453**, 80–88 (2008)
5. Kim, H., Sah, P., Yang, C., Roska, T., Chua, L.O.: Memristor bridge synapses. Proc. IEEE **100**(6), 2061–2070 (2012)
6. Pershin, Y.V., Di Ventra, M.: Practical approach to programmable analog circuits with memristors. IEEE Trans. Circ. Syst. I **57**(8), 1857–1864 (2010)
7. Domnguez-Castro, R., Espejo, S., Rodrguez-Vzquez, A., Carmona, R.A., Fldesy, R., Zanrdy, A., Szolgay, P., Szirinyi, T., Roska, T.: A 0.8-m CMOS two-dimensional pro-grammable mixed-signal focal-plane array processor with on-chip binary imaging and in-structions storage. IEEE J. Solid-State Circ. **32**, 1013–1026 (1997)
8. Kim, H., Sah, M.P., Yang, C., Roska, T., Chua, L.O.: Neural synaptic weighing with a pulse-based memristor circuit. IEEE Trans. Circ. Syst. **I**(59), 148–158 (2012)
9. Guide, Virtuoso Spectre Circuit Simulator User: CADENCE. San Jose, CA, USA (2004)
10. Wolfram, S.: Universality and complexity in cellular automata. Phys. D Nonlinear Phenom. **10**(1), 1–35 (1984)
11. Itoh, M., Chua, L.O.: Memristor cellular automata and memristor discrete-time cellular neural networks. Int. J. Bifurcat. Chaos **19**(11), 3605–3656 (2009)
12. Chua, L.O.: Memristor-the missing circuit element. IEEE Trans. Circ. Theory **18**(5), 507–519 (1971)
13. Ascoli, A., Corinto, F., Tetzlaff, R.: Generalized boundary condition memristor model. Int. J. Circ. Theory Appl. (2015)
14. Orlowski, M., Secco, J., Corinto, F. Chuas constitutive memristor relations for physical phenomena at metal-oxide interfaces. J. Emerg. Sel. Top. Circ. Syst. (2015). (in press)
15. Baldassi, C., Braunstein, A., Brunel, N., Zecchina, R.: Efficient supervised learning in networks with binary synapses. BMC Neurosci. **8**(Suppl 2), S13 (2007)

Reinterpretation of Magnetic Tunnel Junctions as Stochastic Memristive Devices

Adrien F. Vincent, Nicolas Locatelli and Damien Querlioz

Abstract Spin-transfer torque magnetic random access memory (STT-MRAM) is currently under intense academic and industrial development, since it features non-volatility, high write and read speed, and outstanding endurance. The basic cell of STT-MRAM, the spin-transfer torque magnetic tunnel junction (STT-MTJ), is a resistive memory that can be switched by electrical current. STT-MTJs are nevertheless usually not considered as memristors as they feature only two stable memory states. Their specific stochastic behavior, however, can be particularly interesting for synaptic applications and can allow us reinterpreting STT-MTJs as "stochastic memristive devices." In this chapter, we introduce basic concepts relating to STT-MTJs behavior and their possible use to implement learning-capable synapses. Using system-level simulations of an example of neuroinspired architecture, we highlight the potential of this technology for learning systems. We also compare the different programming regimes of STT-MTJs with regard to learning and evaluate the robustness of a learning system based on STT-MTJs to device variations and imperfections. These results open the way for unexplored applications of magnetic memory in low-power, cognitive-type systems.

1 Introduction

Spintronic devices constitute an emerging class of electron devices where the intrinsic magnetic moment of the electrons—their spin—plays a major role, in addition to their electrical charge. In recent years, a flagship application of spintronics, magnetic random access memory (MRAM), has appeared as a breakthrough for nonvolatile

A.F. Vincent (✉) · N. Locatelli · D. Querlioz
Institut d'Électronique Fondamentale, Université Paris-Sud, Orsay, France
e-mail: adrien.vincent@u-psud.fr

N. Locatelli
e-mail: nicolas.locatelli@u-psud.fr

D. Querlioz
e-mail: damien.querlioz@u-psud.fr

© Springer (India) Pvt. Ltd. 2017

M. Suri (ed.), *Advances in Neuromorphic Hardware Exploiting Emerging Nanoscale Devices*, Cognitive Systems Monographs 31,
DOI 10.1007/978-81-322-3703-7_5

memory chips. More precisely, spin-transfer torque magnetic random access memory (STT-MRAM), the second generation of MRAM, provides a combination of nonvolatility and fast programming. These features are shared with several emerging nonvolatile memory technologies, which do not make use of spintronics. STT-MRAMs, by contrast, also provide outstanding endurance [13, 53], a much rarer feature, and are in the process of industrialization by several major companies. The unique properties of STT-MRAM are the result of important progress made in recent years on their basic cell, a resistive switching element known as the spin-transfer torque magnetic tunnel junction (STT-MTJ). Yet, a drawback of this technology remains: The switching between the memory states of STT-MTJs is of a stochastic nature [13, 26, 55]. The time needed for programming from a memory state to another is a random quantity, and this phenomenon came under physicists scrutiny [6, 12]. In conventional memory applications, this drawback requires designing programming pulses long enough to ensure high enough safety margins. Elaborate circuit concepts, as self-enabled programming, have been proposed by circuit designers to mitigate this issue [26]. However, an alternative point of view can be to consider this source of randomness as a *feature*, and not as a weak point of the device.

In particular in this chapter, we reinterpret STT-MTJs' behavior as a "stochastic memristive device" and show a way of how it may be used in a neuromorphic system for practical applications. Numerous works have proposed to use memristive devices as synapses in neuromorphic systems [18, 19, 36, 39, 42]. Usually, they rely on multilevel memory devices, close to the original memristor paradigm [43]. An alternative idea is to use binary devices programmed in a stochastic fashion, or even to use binary devices with intrinsic stochastic properties [15, 21, 48, 49], as has also been proposed in theoretical works [25, 28, 40]. We suggest that STT-MTJs are ideal for this vision and illustrate it in the case of unsupervised learning.

In the present chapter, we introduce the basic physics of STT-MTJs and the foundations of their behavior as stochastic memristive devices. We illustrate this last idea with system-level simulations incorporating an accurate model of the stochastic effects in the case of a practical application of car counting. Monte Carlo simulations show the relevance and the robustness of the approach to device variations and are used to identify in which regime STT-MTJs should preferentially be used.

2 Magnetic Tunnel Junction Basics

2.1 *Basic Structure of Magnetic Tunnel Junctions*

The basic structure of a magnetic tunnel junction (MTJ) is composed of two ferromagnetic layers, surrounding an insulator layer. This last layer is thin enough (0.5 to 2 nm) to allow a tunneling current to flow through the device. As presented in Fig. 1, the magnetic moment of the so-called reference layer is pinned in a fixed direction (usually by means of additional layers that are not represented here), while the mag-

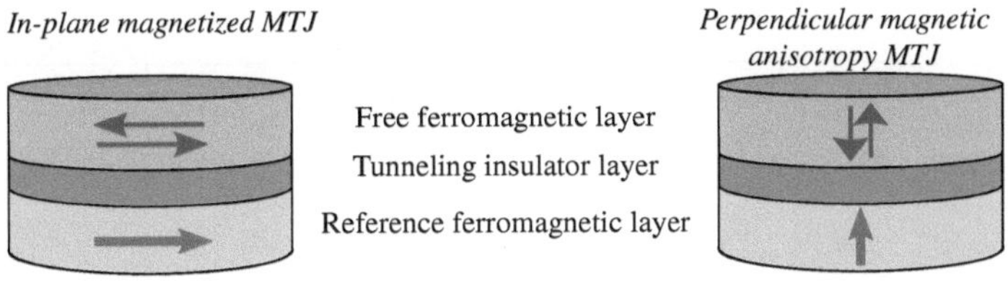

Fig. 1 The principal layers in magnetic tunnel junctions. The arrows represent the magnetization of the different layers. *Left* a configuration with an in-plane magnetization. *Right* a device with perpendicular magnetization

netic moment of the "free layer" can adopt two different relative orientations with respect to the reference layer, parallel (P) or antiparallel (AP). Furthermore, the two ferromagnetic layers can be magnetized either in-plane or out-of-plane, depending on the dominant anisotropy axis.

Due to the spin-dependent electron tunneling, the electrical resistance of MTJs depends on the relative orientation of the ferromagnetic layers' magnetizations. Assuming that the spins of the electrons are conserved during the tunneling, one can consider that the total electronic current flowing through an MTJ results from the contributions of two independent parallel channels, as electrons' spins can have two different directions. In a ferromagnet, the existence of a nonzero magnetization is associated with a dissymmetry in the density of electrons depending on their spin direction. In an easy picture, electrons whose magnetic moment direction (related to their spin) is identical to the local magnetization are majority, when electrons whose magnetic moment direction is opposite are minority. Identically, when tunneling into a ferromagnetic layer, an incoming electron whose magnetic moment is opposite to the local magnetization will have a higher probability to be reflected than an electron whose magnetic moment is parallel to the local magnetization. In an MTJ, when the magnetizations of both ferromagnetic layers are in the same direction (parallel), the majority spin electrons emitted by the first ferromagnetic layer (FM 1) will have a higher tunneling probability into FM 2 than the minority spin electrons. In the other case, when the magnetizations are antiparallel, the minority electrons from FM 1 have a high tunneling probability into FM 2, while the majority spins have a low tunneling probability. The resulting electronic current is then lower than in the first case, the resistance is higher. This simple model explains why the resistance R_{AP} in the antiparallel state is higher than the resistance R_P in the parallel state [20].

The *tunnel magnetoresistance* (TMR) is a metric used to characterize the amplitude of this phenomenon and is defined as

$$\text{TMR} = \frac{R_{AP} - R_P}{R_P}. \tag{1}$$

Higher TMR values mean higher separation between the AP and P states, in terms of resistance values. TMR in mature technologies typically ranges 100–200 % and can reach 600 % in academic technologies. It should be noted that the TMR is voltage

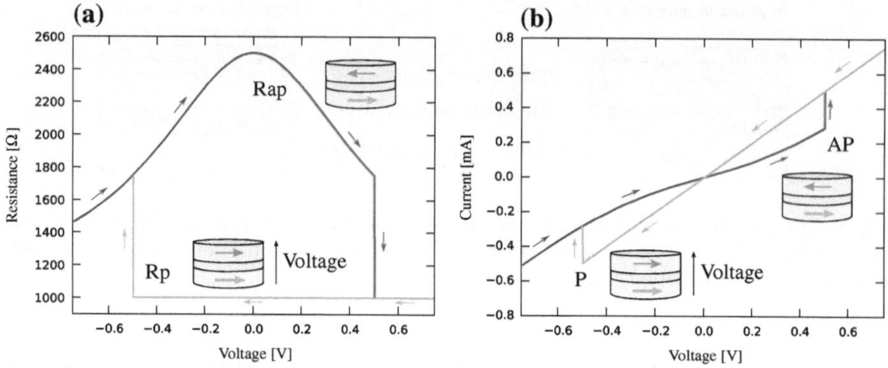

Fig. 2 Sketch of the resistance voltage **a** and current voltage **b** characteristics of an MTJ

dependent, because of the evolution of the resistance in the antiparallel state R_{AP} with respect to the voltage. A sketch of this dependence is given in Fig. 2a. Figure 2b is the sketch of the equivalent current-voltage characteristic.

We can also see on these curves that it is possible to switch the state of the MTJ with electrical currents: High positive currents can switch MTJ from the AP to P state, while high negative currents can switch it from the P to AP state. The detailed physics (spin-transfer torque) and behaviors associated with MTJ resistive switching are explained in Sect. 2.4, but we can already remark that the MTJ is reminiscent of a bipolar memristive device. Its two distinctive features are that MTJ are true binary devices (no intermediate state between P and AP is stable), and the stochastic nature of switching, largely discussed in Sect. 2.4.

2.2 Integration and Scaling Potential of STT-MTJs

In terms of materials, the ferromagnetic layers are usually made of CoFeB alloy. The barrier layers were originally made of aluminum oxide, but are now replaced by crystalline MgO to maximize the TMR values. The reference CoFeB layer is usually coupled to a third CoFe ferromagnet in a CoFeB/Ru/CoFe structure constituting a "synthetic antiferromagnet" (SAF) in order to cancel the reference layer's dipolar influence on the free layer. This SAF is then pinned by the addition of a PtMn antiferromagnet, completing the standard MTJ stack.

One of the great advantages of MTJs is their back-end-of-line (BEOL) compatibility with a standard CMOS process, as they can sustain annealing of 350 °C [35]. From a technological point of view, the most challenging step is the etching of the individual MTJs.

The integration potential of STT-MTJ has been demonstrated in many realizations, which use STT-MTJs as standard memory. A standalone memory of 64 Mb, in a

90 nm process, has already reached the market [2, 38], and similar chips have been published by several groups [5, 11, 54]. An embedded memory of 1 Mb in a 65 nm process has been published [34].

Unlike flash memory, as well as several alternative nonvolatile memory technologies, STT-MTJs use programming voltages lesser or equal than logic voltages. Programming currents scale with the technology node [13] and can range from mA to 10 μA. The most recent realizations use structures with out-of-plane magnetization (PMA MTJs, referring to the perpendicular magnetic anisotropy), which reduces the programming current [23, 24, 50]. In Ref. [24, 34], with 30 nm PMA MTJs, programming voltage is 0.6 V, programming current is 50 μA and programming time is only 3 ns. The read and write circuits associated with STT-MTJs have also been heavily developed in recent years. Advanced read circuits make use of sense amplifiers specially designed for STT-MTJs [34, 56], while advanced write circuits mitigate stochastic effects using self-enabled paradigms [26].

2.3 *Physical Modeling of Magnetization Dynamics*

A usual approximation to describe the dynamics of an MTJ is to assume that the total magnetic moment of the free layer behaves as a single "macropsin" $\mathbf{m_f}$, with a coherent reversal process between the two stable directions. This approximation is believed to capture the essential behavior of in-plane MTJs. With perpendicularly magnetized MTJs, macrospin predictions tend to become less accurate when the devices lateral size exceeds ~50 nm. Beyond this boundary, one should include subvolume thermal magnetic fluctuations to account for the experimental evidences demonstrating the failure of the macrospin reversal model [46].

The precessional motion of the macrospin $\mathbf{m_f}$ is well described by the Landau–Lifshitz–Gilbert equation, in which additional terms are added to account for the thermal noise, and for the spin-transfer torque (STT):

$$\frac{d\mathbf{m_f}}{dt} = \underbrace{-|\gamma|\mu_0 \mathbf{m_f} \times (\mathbf{H}_{\text{eff}} + \mathbf{h}_{\text{sto}})}_{\text{Precession}} + \underbrace{\frac{\alpha}{M_s V} \mathbf{m_f} \times \frac{d\mathbf{m_f}}{dt}}_{\text{Damping}} + \underbrace{V\mathbf{T}_S}_{\text{STT}} \tag{2}$$

(all the equations of the present chapter are given in SI units.)

The effective field $\mathbf{H}_{\text{eff}}$ includes the different magnetic anisotropy terms and a possibly applied external field. The parameters γ, μ_0, and V are, respectively, the electron gyromagnetic ratio, the vacuum magnetic permeability, and the volume of the free layer. The material-related parameters are the saturation magnetization M_s and the dimensionless Gilbert damping parameter α. $\mathbf{T}_S$ stands for Slonczewski's spin-transfer torque term [14, 41] and is current dependent. The smaller field-like STT term is here neglected. The stochastic Langevin term $\mathbf{h}_{\text{sto}}$ models the thermal effects: Its Cartesian coordinates are assumed to be independent Gaussian stochastic processes with zero average and no correlation [14, 16].

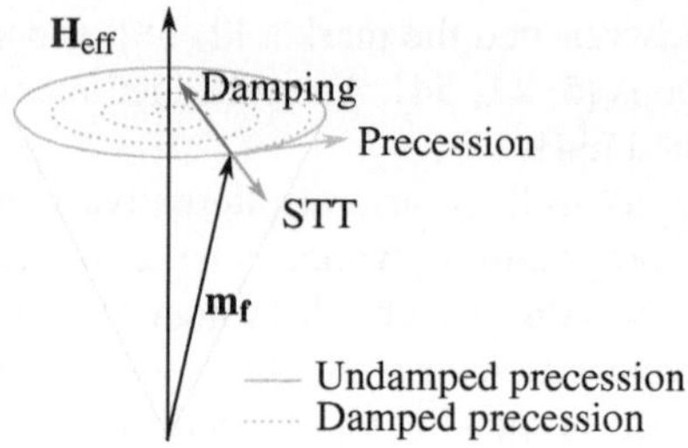

Fig. 3 Sketch of the trajectory of a magnetic moment $\mathbf{m_f}$ around an effective field $\mathbf{H_{eff}}$ (for clarity, $\mathbf{h_{sto}}$ is presently assumed to be zero), without damping (*solid ellipse*) or with damping (*dotted spiral*). The *three colored arrows* represent the three torques' components appearing in Eq. (2). The STT torque may be in the opposite direction, depending on the sign of the current

Figure 3 gives a sketch of the effect of each of the three different components that are present in Eq. (2):

- the precession torque, because of which the magnetic moment $\mathbf{m_f}$ tends to describe a cyclic trajectory around the effective field $\mathbf{H_{eff}}$;
- the damping torque, characterized through the α factor, which makes the magnetic moment $\mathbf{m_f}$ relax along the direction of the effective field $\mathbf{H_{eff}}$;
- the spin-transfer torque, which can act as an antidamping component or an over-damping component, depending on the sign of the current flowing through the MTJ.

The spin-transfer torque effect is the physical effect which allows to act on the magnetization of the MTJ's free layer by simply flowing current through it. If the injected current is of the correct direction and of a sufficient magnitude to overcome damping, the magnetic moment $\mathbf{m_f}$ can switch into the other configuration.

A phenomenological explanation of the spin-transfer torque effect is sketched in Fig. 4. Each electron possesses a magnetic moment, induced by its spin angular momentum. For positive currents, the magnetic moments of conduction electrons tend to align with the local magnetization of the first ferromagnetic layer that they encounter, which is the reference layer: This layer acts as a spin polarizer. During the tunneling through the insulator layer, the spin polarization is conserved and the second ferromagnetic layer, the free layer, sees an incoming flow of polarized electrons. As they flow through the free layer, the magnetic moments of these electrons lose their transverse component to align along the free layer's magnetization. Considering spin/momentum conservation, this component is actually transferred to the free layer, resulting in a torque acting to tilt its magnetization, as pictured in Fig. 4a. This phenomenon is thus called *spin-transfer torque* (STT), and its magnitude is proportional to the flow of electrons, hence the current. When the current is reversed, the free layer is still subject to a spin-transfer torque, of opposite direction, due to electrons that are reflected back from the reference layer, as cartooned in Fig. 4b. One can find more details about the spin-transfer phenomenon in the reference [37].

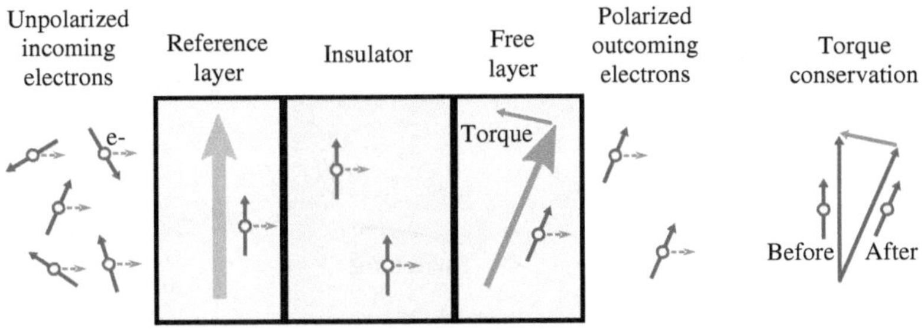

(a) Spin-transfer torque on the free ferromagnetic layer, due to the crossing electrons (positive current).

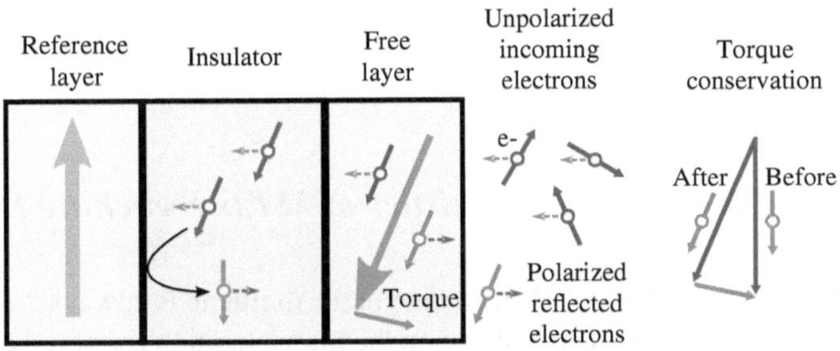

(b) Spin-transfer torque on the free ferromagnetic layer, due to the reflected electrons (negative current).

Fig. 4 **a** For positive currents, the first ferromagnetic layer crossed by the incoming electrons is the reference layer and acts as a polarizer. After tunneling, the electrons realign their magnetic moment along the magnetization of the free layer, which results in applying a spin-transfer torque (STT) to it (see sketch on the *right* part of the figure). **b** When the current is reversed, the spin-transfer torque acting on the free layer is due to electrons reflected from the reference layer. The torque direction is then reversed

With the correct sign and a sufficient magnitude of current, the spin-transfer torque can overcome the damping torque and make the magnetization $\mathbf{m_f}$ of the free layer switch into the other direction.[1]

An illustration of the reversal process is plotted in Fig. 5. The trajectory of the free layer's magnetization (blue solid line) was obtained by numerically solving Eq. (2) in the presence of a DC current. The spiral oscillations of the trajectory are the result of the precession component; the trajectory is jittered as the magnetic moment is subject to thermal agitation. The magnetic moment $\mathbf{m_f}$ is progressively dragged out of its initial position, before switching into the opposite direction. After reversing, the DC current does not act as an antidamping effect any more, but as a supplementary damping torque; $\mathbf{m_f}$ then converges quickly toward its new equilibrium position.

[1]It should be noted that the reference layer also experiences spin torque, but cannot switch as it is intentionally pinned.

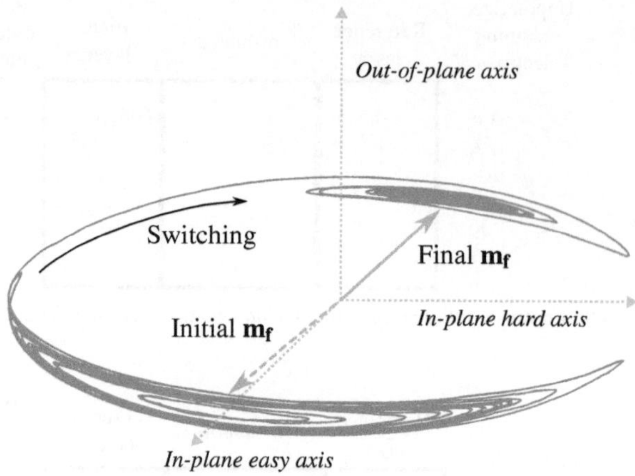

Fig. 5 Example of one trajectory of the magnetic moment $\mathbf{m_f}$ that is given by Eq. (2) for an in-plane magnetized MTJ, when a DC current is injected and the temperature is nonzero

2.4 Models About the Statistics of MTJs Switching Delay

Except the case at $T = 0$ K, a magnetic moment is always subject to thermal agitation. The impact of such noise on the magnetization $\mathbf{m_f}$ of the free layer is taken into account in Eq. (2) by adding a Langevin term $\mathbf{h}_{\text{sto}}$. Mathematical details on this particular term can be found in the reference [16]. The important consequence of this observation is that for an MTJ at room temperature, the switching process will *always* be affected by a stochastic component. The switching delay—between the instant when a current or a voltage is applied, and the instant when the magnetization $\mathbf{m_f}$ has switched—is thus intrinsically a random quantity [6, 12, 13].

Different regimes of stochasticity of the switching delay exist, depending on the programming current amplitude, which determines the magnitude of the spin-transfer torque. In the next subsections, we identify three different regimes and briefly present how they differ from one another and how one can model them. It is an important step to further be able to exploit this intrinsic stochasticity of MTJs as a feature in neuromorphic architectures.

In the following, we will focus on a study of in-plane magnetized STT-MTJs and assume that no external magnetic field is applied on them.

2.4.1 A Critical Value of Current Density for Magnetization Reversal

For a correct choice of the current sign, the spin-transfer torque has an antidamping effect, as explained previously in Sect. 2.3. The critical current density magnitude J_{c0} can be defined as the limit of stability of the free layer's magnetization when current is injected at $T = 0\,\text{K}$ [45].

Physically, putting aside the thermal fluctuations and considering a slightly out-of-equilibrium magnetic moment $\mathbf{m_f}$, one can mainly encounter two situations, depending on the magnitude of the injected current density J_s:

$J_s < J_{c0}$: The damping torque is bigger than the spin-transfer torque and the magnetic moment $\mathbf{m_f}$ relaxes into its initial position (as the dotted trajectory in Fig. 4a).

$J_s > J_{c0}$: The spin-transfer torque overcompensates the damping torque, destabilizing the equilibrium state. The precession radius of $\mathbf{m_f}$ increases until $\mathbf{m_f}$ switches in the opposite configuration.

This critical current density magnitude can be expressed, in the case of an in-plane magnetized MTJ [44, 45], as

$$J_{c0} = \frac{2|e|}{\hbar} \times \frac{1 \pm P^2}{P} \times \alpha t_f \mu_0 M_s \left(H_k + \frac{M_s}{2} \right), \tag{3}$$

where $|e|$ and $\hbar$ are, respectively, the elementary charge and the reduced Planck constant. The $\pm$ sign should be $-$ for the AP→P transition, and $+$ for the P→AP transition. The term P is the spin polarization of the current (which can be computed from the TMR using $\text{TMR} = 2P^2/(1-P^2)$) and H_k is the amplitude of the anisotropy field. This anisotropy field can result from the summation of different anisotropies contributions. It may, for example, be the combination of material-related magnetic anisotropy, such as crystalline anisotropy, and of shape anisotropy that results from the device geometry (in case of an elliptical cross section for instance).

The corresponding equation in the case of out-of-plane magnetization can be found in [22].

The spin torque amplitude appears not to be equivalent for both signs of the current, so that J_{c0} has two different values, depending on the transition that is considered (P→AP or AP→P). However, basic calculations, confirmed by measurements, show that they correspond to the same critical voltage: $J_{c0}^{P\rightarrow AP} R_P = J_{c0}^{AP\rightarrow P} R_{AP} = Vc$ (as illustrated in Fig. 2). This symmetry is an interesting property of STT-MTJs when used as memristive devices.

In Fig. 6, dots show the evolution of the mean switching times obtained from solving Eq. (2) in macrospin-based Monte Carlo numerical simulations, at 300 K.

> To have an idea of the order of magnitude of the critical current density, let us consider the case of an MTJ with $P = 0.65$ (which corresponds to a TMR around 150 %), typical values for α, t_f, and M_s of, respectively, 0.01, 2 nm, and $1.2 \times 10^6 \text{A} \cdot \text{m}^{-1}$. The magnitude of the anisotropy field H_k is usually smaller than the saturation magnetization M_s. We will assume that $H_k \sim \frac{1}{10} M_s$. In this situation, Eq. (3) results in $J_{c0} \sim 9 \times 10^6 \text{A} \cdot \text{cm}^{-2}$ for an AP→P event, and $J_{c0} \sim 22 \times 10^6 \text{A} \cdot \text{cm}^{-2}$ for a P→AP event.

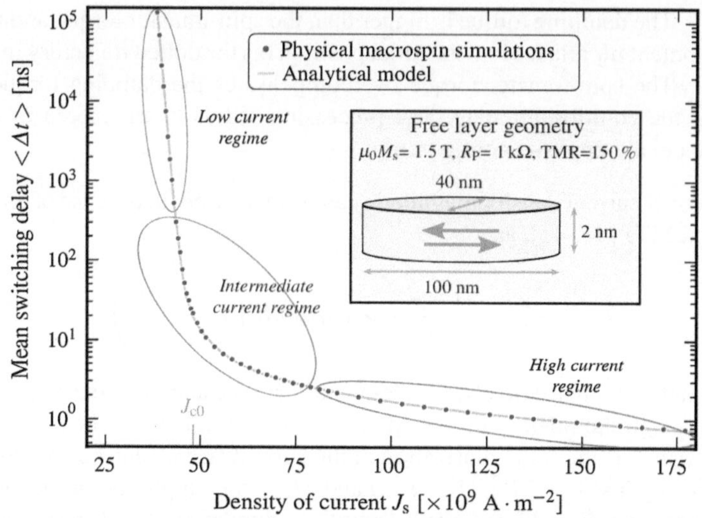

Fig. 6 Evolution of the mean switching delay $< \Delta t >$ with respect to the injected current density J_s, for an AP → P switching event. The *blue dots* are the average values on Monte Carlo simulations of Eq (2). The *green solid line* is the prediction coming from analytical equations detailed in the reference [52]. *Orange ellipses* identify the three regions with a different behavior of $< \Delta t >$. Some information about the simulated free layer is given in the inset, and the critical current density J_{c0} is indicated on the x-axis

2.4.2 In the Low-Current Regime

When the magnitude of the current density J_s that flows through a STT-MTJ is significantly lower than its critical value J_{c0}, the reversal process of the magnetic moment $\mathbf{m_f}$ is dominated by the thermal activation.

Solving the Fokker–Planck equation derived from Eq. (2) results in the Néel–Brown model [3, 10, 29, 32] of the STT-MTJ mean switching delay

$$< \Delta t > = f_0^{-1} \times \exp\left(\frac{E_0}{k_B T}\left(1 - \frac{J_s}{J_{c0}}\right)\right), \tag{4}$$

where $E_0 = \mu_0 M_s H_k V/2$ is the energy barrier separating the two stable states for the magnetization at zero current, and $k_B T$ is the thermal energy. The prefactor f_0^{-1} is a time constant, usually chosen of the order of 1 ns; it is related to the natural precession timescale [3, 44].

To illustrate how the mean switching delay $<\Delta t>$ can easily be tuned over several decades in the low-current regime, let us assume (for example) an energy barrier $E_0 = 40\ k_B T$ (which in average retains a given state during approximately 7.5 years with $f_0^{-1} = 1$ ns). In that case, the model of Néel–Brown predicts a value of $<\Delta t>$ about 0.2 µs when $J_s = \frac{3}{4} J_{c0}$. If one now reduces the current by one-third (i.e., $J_s = \frac{1}{2} J_{c0}$), then $<\Delta t>$ almost reaches 0.5 s.

One can effectively observe in Fig. 6 that, in the low-current regime, the mean switching delay $<\Delta t>$ exponentially decreases with respect to the current density J_s, which agrees with the predictions of the model of Néel–Brown.

In this regime, the injected current thus acts like it increases the effective temperature of the system (or decreases the effective energy barrier), thus increasing the probability to cross the energy barrier [29]. However, it should be noted that this equation was derived in the high-energy-barrier approximation: if the effective energy barrier becomes too small, the Néel–Brown model should thus no longer be valid.

It has been theoretically [29] and experimentally [17] demonstrated that in this regime, the reversal process is a Poisson process, so that the probability density function (PDF) of the switching delay Δt is

$$f_{sw}(t) = \frac{1}{<\Delta t>} \exp\left(-\frac{t}{<\Delta t>}\right), \tag{5}$$

where $<\Delta t>$ is given by Eq. (4): The switching delay Δt follows an exponential random law with mean $<\Delta t>$ given by Eq. (4) [6, 13]. It is what can also be observed in Fig. 7a. With such a probability density function, if a programming pulse of duration t_p (and of constant current) is applied to the junction, its probability of switching is

$$P_{sw} = 1 - \exp\left(-\frac{t_p}{<\Delta t>}\right). \tag{6}$$

In these conditions, choosing the pulse duration t_p allows tuning the switching probability of the devices anywhere between a low ($P_{sw} \ll 1$ for $t_p \ll\ <\Delta t>$) and a high probability ($P_{sw} \approx 1$ for $t_p \gg\ <\Delta t>$).

An important difference with other families of memristive devices is that STT-MTJs possess no hard threshold. Even an extremely low current will increase the probability for the MTJ to switch its state.

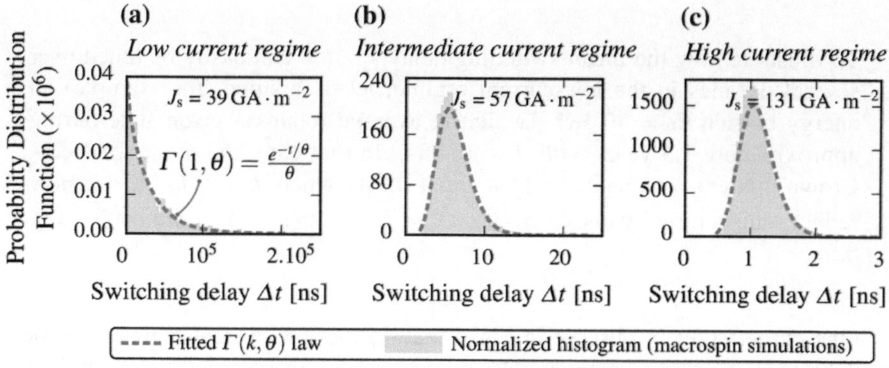

Fig. 7 From **a** to **c**: the probability distribution functions of the switching time Δt in the low-, the intermediate-, and the high-current regimes. The precise value of current density is given in each inset. The *dashed orange lines* compare fitting a two-parameter Gamma law $\Gamma(k, \theta)$ with the normalized histograms (tinted areas) that are computed from the same Monte Carlo simulations as in Fig. 6. In the low-current regime, the exponential law presented in Eq. (5) is recovered by using $k = 1$ and $\theta = <\Delta t>$

2.4.3 In the High-Current Regime

When the current density J_s is of a significantly higher magnitude than J_{c0}, the reversal dynamics becomes dominated by the spin-transfer torque action, which results in a quasi-adiabatic reversal process. In the case of in-plane MTJs, the mean switching delay Δt then behaves according to Sun's law [12, 13, 45]

$$<\Delta t> = \ln\left(\frac{\pi}{2\phi_0}\right) \times \frac{1}{\alpha\gamma\mu_0\left(\frac{M_s}{2} + H_k\right)} \times \frac{J_{c0}}{J_s - J_{c0}}, \tag{7}$$

where ϕ_0 is the standard deviation of the random initial angle of the magnetic moment $\mathbf{m_f}$. The initial angle before reversal is indeed related to thermal fluctuations around the equilibrium position before the current is applied and follows a Gaussian distribution around 0 and a standard deviation $\phi_0 = \sqrt{k_B T/(\mu_0 H_k M_s V)}$. Sun's model then considers that the reversal trajectory is not affected by thermal fluctuations. It is then often referred to as "precessional switching." This model is thus no longer valid when the current density is low enough for the Brownian motion of $\mathbf{m_f}$ to notably affect its trajectory.

The corresponding equation in the case of out-of-plane magnetization can be found in [22].

If we consider an MTJ with the same parameters as for the previous examples and if we assume a plausible value of 4° for ϕ_0, then for $J_s = 2\ J_{c0}$ the mean switching delay is about 3ns. In the high-current regime, the switching delay is thus much faster in average than in the low-current regime. It is, for example, the targeted programming regime in memory chips.

In this regime, the probability density function of $< \Delta t >$ is no longer an exponential law as is seen in Fig. 7c. One can instead use a Gamma law $\Gamma(k, \theta)$ to model the probability distribution of the switching delay. Details are given in the reference [52].

2.4.4 The Intermediate-Current Regime

Between the low-current and high-current regimes, thermal fluctuations can no longer be neglected during the switching process, although the spin-transfer torque also plays a significant role. Since none of the effects can be neglected in this intermediate regime, an analytical expression of the mean switching delay cannot be easily derived from the dynamics equation. The non-negligible influence of the thermal noise on the switching trajectory makes the dynamic harder to model than in the two other regimes [14]. Furthermore, there is no easy connection between the two extreme models, since one can notice that according to Sun's model the mean switching delay diverges for $J_s = J_{c0}$.

However, one can use the comprehensive analytical model from reference [52]. The latter is compared, in the case of an in-plane magnetized MTJ, with the physical macrospin Monte Carlo simulations in Figs. 6 and 7b, showing good agreement.

3 MTJs as Stochastic Synapses

Artificial neural networks usually rely on synapses that can feature multiple conductance values (synaptic weights) to achieve learning. However, it has been suggested that a neural network with binary weight synapses is still able to perform learning if one adopts a probabilistic learning rule, in particular for supervised learning [25, 40].

Let us first consider a conventional learning-capable system with multiple-weight synapses and a deterministic learning rule. During each learning step, that occurs after an input pattern is presented, every relevant synapse increases or decreases its conductance of a small quantity. The combination of all these small variations results in a global change of the network.

By contrast, in a system with binary synapses, the synapses' conductance can only evolve with an "all or nothing" fashion. A way to achieve similar global change as in a multiple-weights synapses system is that, during each learning step, only *a small fraction* of the synapses whose weight should evolve according to the learning rule actually do, while the weight of the others—most synapses—does not evolve at all. A proposed manner to achieve this is to implement a probabilistic learning rule: during a learning step, each synaptic switching from one weight to another that is supposed to occur only has a limited *probability* to happen.

Some work recently successfully demonstrated the relevance of this approach in the case of unsupervised learning by applying a stochastic learning rule to binary CBRAM synapses [49]. In this work, an external pseudorandom number generator (PRNG) is used to probabilistically program the deterministic synaptic devices.

A peculiar feature of MTJs-based synapses is that one has a good understanding of the intrinsic randomness of their switching delay. It is therefore possible to exploit this behavior to directly implement the aforementioned stochastic learning rule, without using an external PRNG.

In the current section, we will present the design and the simulation of a neuromorphic system with synapses made of a single MTJ and that are probabilistically programmed by harnessing the randomness of their switching delay. This example will be used to discuss how to program these binary stochastic MTJ-based synapses in a way that ensures both a high resilience to device variations and a low programming power. Extensive details about the system itself or the numerical results can be found in the reference [51], which the current section is based on. The simulated system adapts a scheme that was originally proposed for phase-change memory [9, 47] and conductive bridge RAM (CBRAM) [48, 49] synapses. This example of neuromorphic system uses a simplified version of the spike-timing-dependent plasticity (STDP) learning rule, adapted to stochastic binary synapses. STDP is a model of biological plasticity [7, 31] and has been used widely as an inspiration for works using memristive devices as synapses [19, 39, 42]. Here, we abstract the STDP biological models, by making major simplifications [36] which make its implementation natural with STT-MTJs.

3.1 Example of a Feed-Forward Spiking Neural Network Using MTJ-based Synapses

3.1.1 Architecture and Operation of the System

The system implements a spiking neural network: The computation units (CMOS neurons) communicate by asynchronous spikes, similarly to biological neurons. Figure 8 shows the basic architecture of the system. CMOS input neurons present spikes, which may come directly from a neuromorphic sensor. For example, in the presented case, each input neuron corresponds to one pixel of a bioinspired silicon retina filming a 6-lane freeway [1, 30]. The system uses a properly chosen number of CMOS output neurons. The STT-MTJs are organized as a crossbar-connecting input and output neurons in an "all-to-all" manner.

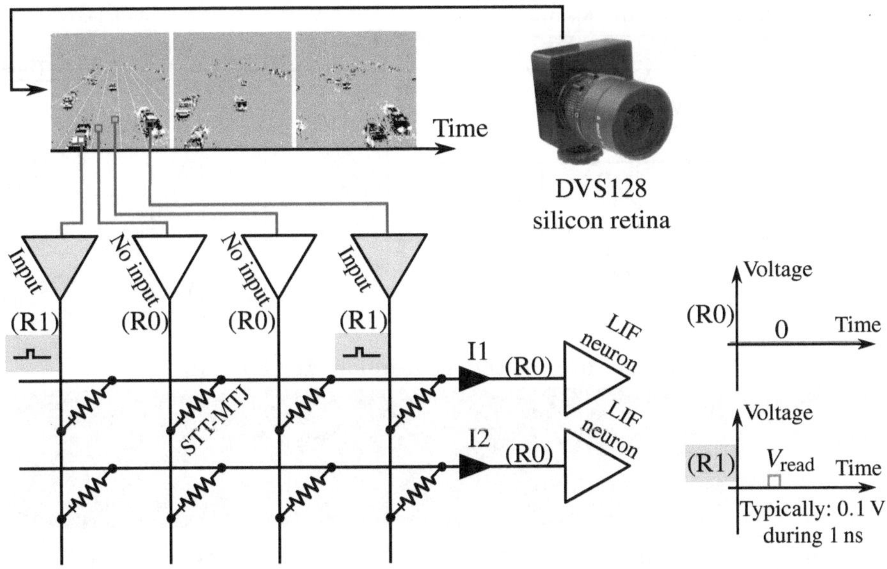

Fig. 8 Cartoon of the 1R crossbar architecture for the learning system, during read operation, which occurs whenever an input neuron spikes and applies a (R1) waveform. The two types of concurrently used voltage waveforms are sketched on the *right*. Each input neuron (▽) is connected to every output neuron (▷) through a STT-MTJ synapse. At the *top* of the figure, one can see some frames made from the input spikes: vehicles are driving toward the DVS retina, on six different lanes (*dotted yellow lines*)

For simplicity, the system is here presented with STT-MTJs organized in a passive crossbar. Physical implementation might require the use of selector devices (1T-1R structure), although this loses the compactness of the scheme, to limit leakages.

From the operation point of view, when an input neuron (▽) spikes, it applies a brief read pulse (R1) to the crossbar, while output neurons (▷) maintain a constantly null voltage (R0) at their input, as illustrated in Fig. 8, while reading the current.[2] Induced current reaches the different output neurons simultaneously. The current received by each output neuron depends on the state (P or AP) of the synapse that connects the input to this particular output. Functionally, the output neurons implement leaky integrate-and-fire (LIF) spiking neurons, a simple and standard neuronal model. Due to design choices that are detailed in [51], only the input spikes going through synapses in the parallel state are integrated by the output neurons. Besides, when an output neuron spikes, it inhibits the other output neurons, resetting their internal variable to zero.[3] The architecture is therefore reminiscent of a "winner-takes-all" structure, widely used in the field of neural networks.

[2]This can be achieved using second-generation current conveyor designs [27].

[3]This can be implemented by nearest-neighbor inhibition [4].

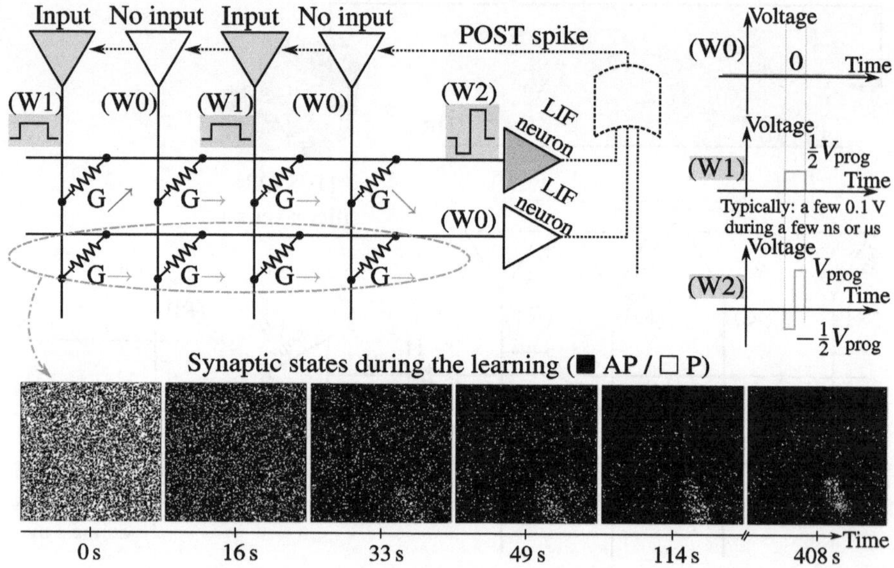

Fig. 9 Cartoon of the 1R crossbar architecture for the learning system, during STDP (write) operation, which occurs when an output neuron spikes. Due to the stochastic nature of switching, in the presented example, only two STT-MTJs switch states (G↗ and G↘) while every other synapses stays in its previous state (G→). *Upper right* sketches of the programming voltage waveforms that are applied concurrently. *Bottom* the evolution during the learning process of the states of the synapses that are connected to one of the output neurons. For visibility, the weights are plotted as a 2-D map

All the simulations' results that are presented in the current Sect. 3 are based on a STT-MTJ device representative of a 45 nm technology. The STT-MTJs are ellipses with a width of 40 nm, a length of 100 nm, and a free layer thickness of 2 nm. The tunnel magnetoresistance (TMR) is 150 % (i.e., $R_{AP}/R_P = 2.5$). The programming voltage amplitude V_{prog} is varied between 0.3 V and 0.6 V, which has an impact on the system operation, as we will see in further sections. The stochastic switching delay is modeled with equations from [52]. More in-depth description and discussion of the system architecture are available in [51].

The stochastic STDP learning rule is implemented by applying concurrent voltage waveforms, as sketched in the upper half of Fig. 9. When an output neuron spikes, the system enters a programming phase[4]: the output neuron that spikes applies a voltage waveform (W2), while only the "recently active" input neurons apply a voltage

[4]Therefore, STDP occurs only when an output neuron spikes. This approach is not common in neurosciences, although Nessler et al. did a similar choice [33].

waveform (W1) at the same time. In the presented case, an input neuron is considered as "recently active" if it spiked during the last 10 ms. This temporal window is very problem dependent, especially on the natural timescale of the presented inputs. As described in detail in Ref. [51], this combination of pulses directly implements the stochastic simplified STDP rule. A STT-MTJ synapse connected to the output neuron that spiked:

- is programmed by a voltage pulse of V_{prog} amplitude and has a given probability $P_{sw}^{AP\to P}$ of switching to the low-resistance P state, if its input neuron was active in a recent time window (given it is not already in the P state);
- is programmed by a voltage pulse of $-V_{prog}$ amplitude and has a given probability $P_{sw}^{P\to AP}$ of switching to the high-resistance AP state if its input neuron was not active in the same time window (given it is not already in the AP state).

At the same time, the STT-MTJs connected to the other output neurons are either nonselected, or half-selected (i.e., the voltage amplitude applied to these devices is either 0 or $V_{prog}/2$). According to the model of Fig. 6 and Eq. (6), the switching probability of these devices is thus negligible if one cleverly chooses $V_{prog}/2$ to correspond to the regime of low programming current.

As the STDP rule is not deterministic but probabilistic, a STDP event has a probability P_{sw} to switch a synaptic nanodevice, as explained in the introduction of the current section.

Assuming that one has a comprehensive analytical model of the probability density function f_{sw} of the switching delay Δt with respect to the voltage amplitude, one can easily design the relevant programming pulses that will make the synapses to switch with the wanted probabilities $P_{sw}^{AP\to P}$ and $P_{sw}^{P\to AP}$. First, setting the pulse amplitude defines the probability density function f_{sw} of Δt. Then, one simply has to tune the pulse duration t_p to precisely ensure the wanted probability to switch

$$P_{sw} = Probability(\Delta t \leq t_p) = \int_0^{t_p} f_{sw}(t)\,\mathrm{d}t. \tag{8}$$

Let us consider again the same MTJ as in the previous examples. Assuming we work in the low-current regime, with a programming pulse corresponding to a current density $J_s = 0.65 J_{c0}$, the Eq. (4) predicts a mean switching time $<\Delta t> = 1.2$ ms. As Eq. (6) is nothing else but the result of Eq. (8), we can derive $t_p = -<\Delta t> \times \ln(1 - P_{sw})$. If, for example, the targeted probability P_{sw} is 10 %, the programming pulse has to last about 125 μs. In the higher-programming-current regimes, solving Eq. (8) often requires numerical techniques, as the analytical expression of the probability density function becomes more complex. In these regimes, typical pulse duration is much more shorter, in the nanosecond range.

3.1.2 Task Results for Car Detection

The neuromorphic DVS retina, used in our example as an input layer, is a sensor that is inspired by the human retina: each input generates spikes when the incoming light intensity of the corresponding pixel changes. The STDP programming pulse width is adjusted such that P → AP and AP → P transitions both have a probability of 10 %. The whole system is simulated by a system-level simulator including a detailed physical model of the STT-MTJs behavior [51].

The system has 20 output neurons and each of them is connected to every input pixel through a synapse made of a single STT-MTJ, as described in Sect. 3.1.1. For each output neuron, one can plot a map of the states (P or AP) of the synapses that are connected to it. The bottom part of Fig. 9 shows some snapshots of such a map for a given output neuron. One can observe how a pattern emerges and stabilizes starting from a random distribution of synaptic states (originally, the STT-MTJ states are random). After 33s of operation, the neuron has started to specialize in lane 4 (see lane labels in Fig. 10). After 114s, the *global* state of the STT-MTJs is stabilized: Although switching events P → AP and AP → P still occur, the pattern does not evolve anymore, even after a significant amount of time.

Therefore, due to the stochastic STDP learning rule, the output neurons naturally specialize on particular lanes and the system effectively becomes a vehicle counter. The specialization of the output neurons is evident in Fig. 10, which shows an overview of the STT-MTJs final states. The top image represents a sample of the inputs: Every input pixel that spiked during a short period is colored with light blue. The yellow dotted lines materialize the six different lanes of the freeway. Every other image represents the final states of the STT-MTJs connected to one of the twenty output neurons (white for P, black for AP). The output neurons are listed according to the lane to which they specialized. The number of output neurons specialized in each lane is not identical. This is determined by the number of cars passing on each lane, and the number of pixels they activate.

To estimate the performance of the system on the car detection task, only the output neuron with the best detection rate is retained for each lane.[5] If one interprets the system as a vehicle counter, the detection rate is 97.3 % (excluding the two outer lanes). The proportion of false positives among the output spikes is 4.7 %.

The best result on the same dataset, using a neural network with double-precision analog weight reports a detection rate of 98.1 % and a proportion of false positives of 4.3 % [8].

Once the learning has been achieved, it is possible to deactivate the learning process to save energy. Besides, the entire system can be switched OFF and ON without losing its function, as the state of STT-MTJs is nonvolatile.

[5] This operation could be done automatically by adding a second layer to the network as proposed in [8].

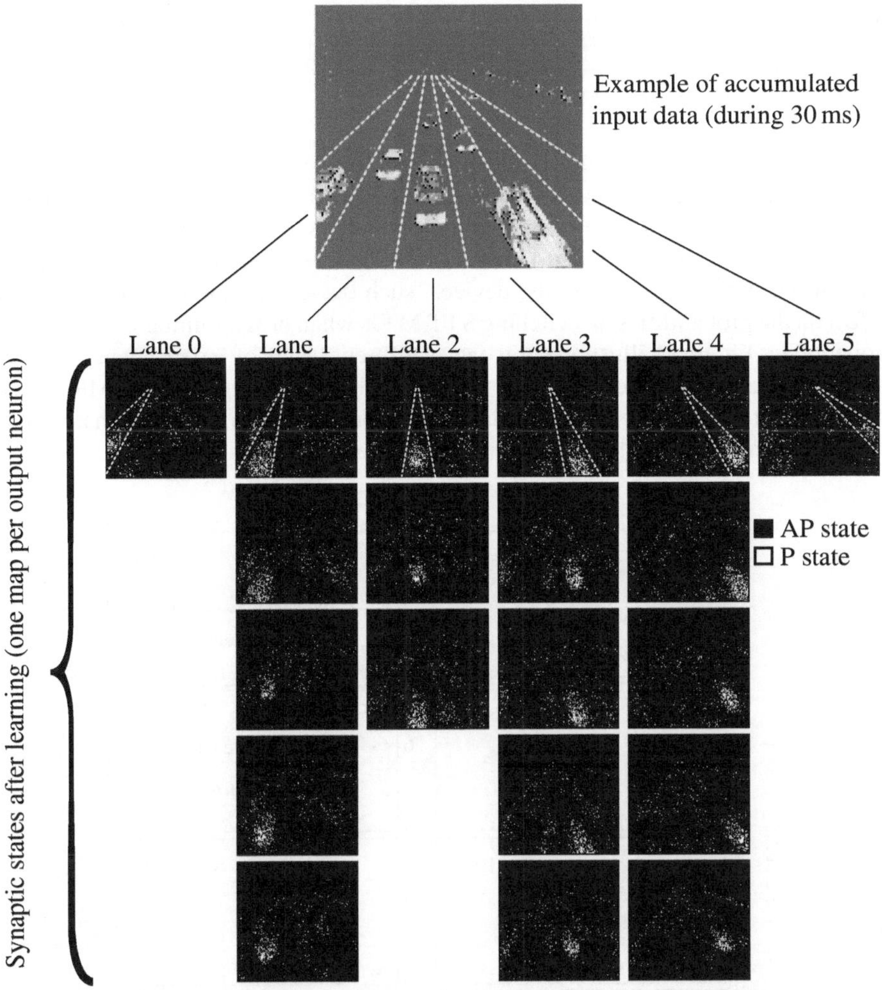

Fig. 10 The *top* image is a sample of the inputs that are presented to the system; the pixels that spiked during the considered 30ms are colored with *light blue*. The other 20 subimages represent the state of the MTJs that are connected to the 20 output neurons (white is P, black is AP), presented as 2-D maps for clarity. The *yellow dotted lines* mark off the driving lanes

It should be noted that, here, a single binary STT-MTJ connects each input to each output. For more complex tasks, it is possible to connect each input to each output by several stochastic STT-MTJs, recreating a multibit synapse [36].

3.2 Impact of the Device Properties on the System Operation

3.2.1 Impact of Device Variations

In a real system, after the shape of the programming pulse is defined (amplitude and duration), the dispersion on the values of the STT-MTJs characteristics will cause the synapses to have different switching probabilities. For instance, let us consider variations of the minimum and maximum resistances of the MTJs. As they affect the current that flows through the devices, such device variations have a dramatic effect on the probabilities of switching STT-MTJs when programming voltage pulses are applied. Figure 11 illustrates this fact. Different levels of independent Gaussian dispersions are introduced on the resistance of the P state and on the TMR. These two parameters have the same standard deviation relative to the mean that will be named "synaptic variability" SV in the rest of the chapter.

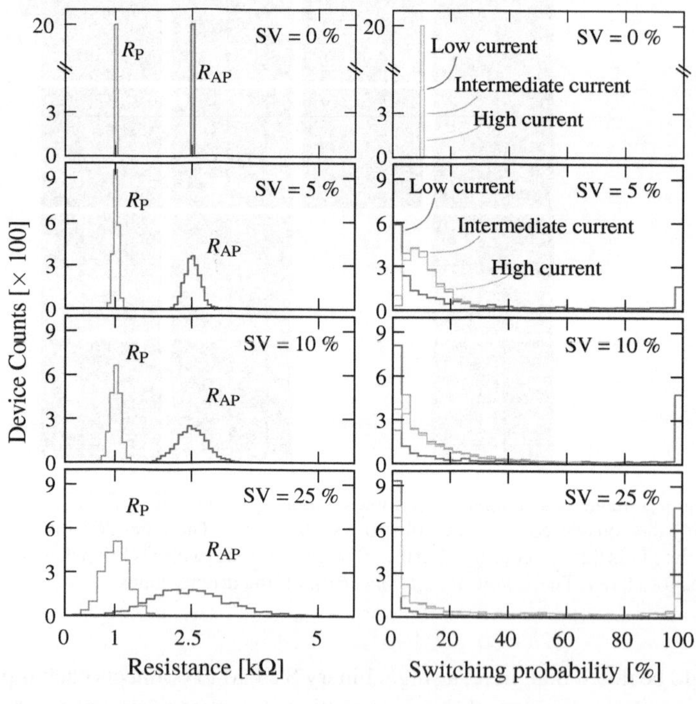

Fig. 11 Histograms representing the values of P and AP states resistance (*left subfigures*) and associated switching probabilities (*right subfigures*), with and without synaptic variability. The switching probabilities are represented in the low-, intermediate-, and high-programming-current regimes, with $V_{prog} = 0.3$ V, 0.4 V, and 0.6 V, respectively. The targeted switching probability P_{sw} is 10 % for all regimes in the case with zero synaptic variability (*top subfigures*). From *top* to *bottom*, synaptic variability SV is 0, 5, 10 and 25 % of standard deviation relative to the mean, on the resistance of the P state and on the TMR

This way of introducing variability is motivated by experimental realizations, which suggest that variations on the resistance of the P state and on the TMR are uncorrelated and have equivalent values of relative standard deviation. In Refs. [5, 54], the SV parameter was found to be approximately 5 %. Most of the synaptic variability values that are further considered—up to 25 %—therefore correspond to extremely high levels of variability, in terms of realistic technology.

The left subfigures of Fig. 11 represent histograms on the resistance values of the parallel and antiparallel states. The right subfigures are computed with the model from [52], mentioned in Sect. 2.4. Given programming pulses designed to ensure a 10 % switching probability for $SV = 0\,\%$, they are the STT-MTJs switching probabilities' histograms once synaptic variability is considered. Three histograms are superimposed in the low-, intermediate-, and high-programming-current regimes. One can observe that the variability on the switching probabilities is exacerbated with regard to the variability on the resistance states. The variability on the switching probabilities is also considerably higher in the low-programming-current regime than in the intermediate- and high-programming-current regimes. One can explain this with Figs. 6 and 7: The distribution of the switching delay Δt in the low-programming-current regime is broader, and its mean value is much more dependent on the current density value (see Eq. (4)) than in the other two regimes.

Yet the whole system reveals to be spectacularly tolerant to device variation. Let us consider the case where the STT-MTJs are programmed in the intermediate-programming-current regime. Figure 12 shows the detection rate and the proportion of false positives as a function of the synaptic variability. No significant impact of the synaptic variability on the detection rate or the proportion of false positives is observed up to high values, largely beyond the ~5 % that are encountered in memory chip demonstrators.

Reference [51] also shows that the system is equally resilient to transient device variations.

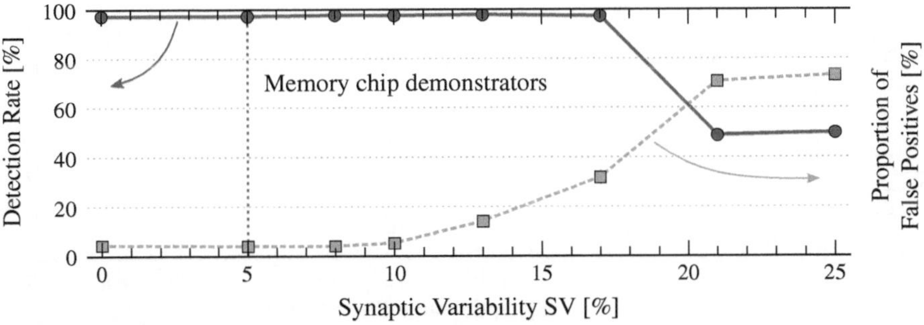

Fig. 12 Detection rate (*blue solid line* and ○) and proportion of false positives (*orange dashed line* and □) as a function of synaptic variability, in the intermediate-programming-current regime ($V_{\mathrm{prog}} = 0.4$ V). Learning and lateral inhibition are disabled at the end of the learning process. Every simulation was repeated ten times

3.2.2 Impact of the Programming Regime

We have previously seen that STT-MTJs can be operated in different programming regimes (low, intermediate, and high current). In this part, we study the benefits and drawbacks of the different regimes when using STT-MTJs as synapses.

Without device variations (SV $= 0\,\%$), one can tailor the programming pulse duration to implement the very same switching probability in all the programming regimes. It therefore leads to the same detection rate and proportion of false positives. However, the average power that is required to program the STT-MTJs differs significantly. For example, during the learning process, the power consumption for programming the STT-MTJs (excluding the power consumption of the CMOS neurons and of the rest of the studied system) can be as low as only a few hundreds of nanowatts in the high-current regime, while it can reach several hundreds of microwatts, or beyond, in the low-current regime, where programming times are longer [51].

In the presence of device variation (SV $\neq 0\,\%$), the programming regimes are furthermore not equivalent in terms of the system operation. This is seen in the Monte Carlo simulations of Fig. 13, where the detection rate and the proportion

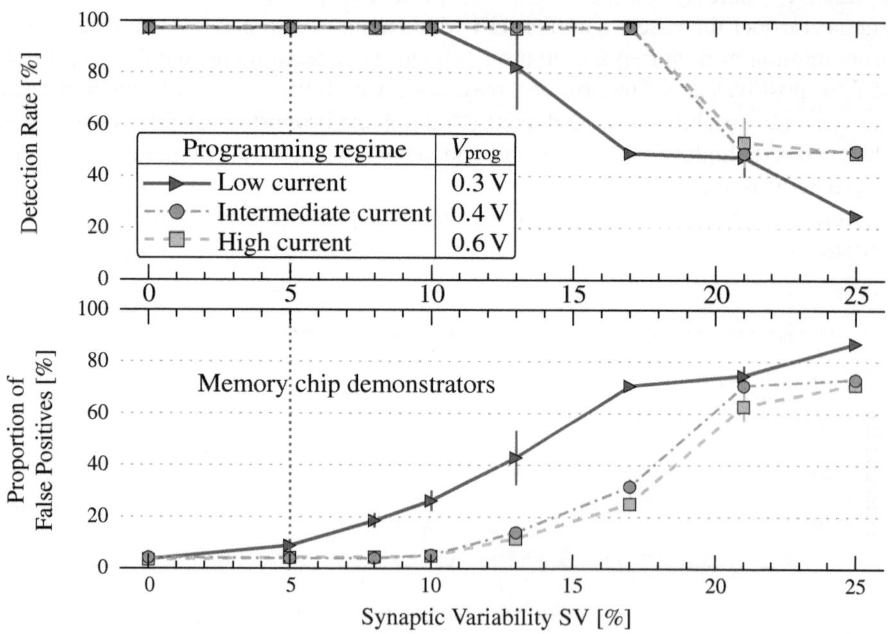

Fig. 13 Detection rate (*top*) and proportion of false positives (*bottom*) as a function of synaptic variability with STT-MTJs programmed in the low (*blue* ▷)-, intermediate (*red* ○)-, or high (*green* □)-programming-current regimes. Both learning and lateral inhibition were disabled at the end of the learning process. Every simulation was repeated ten times, the error bars represent one standard deviation

of false positives are plotted as a function of the synaptic variability, in the three programming regimes. In the low-programming-current regime, the detection rate is observed to start decreasing for values of synaptic variability beyond 10 %, while the number of false-positive events increases as soon as device variations exist. These results are naturally explained by the nonequivalence of programming regimes seen in Fig. 11.

In summary, intermediate- and high-programming-current regimes require smaller programming power than the low-programming-current regime and have a better robustness to device variations. To ensure the highest endurance/reliability for the STT-MTJs, it is preferable using the smallest amplitude possible for the voltage. The intermediate-programming-current regime therefore appears as the ideal regime for synaptic use of STT-MTJs. This differs from more conventional applications where programming speed is the most important, and where high-programming-current might be a preferable choice.

4 Conclusion

In the first part of this chapter, we reviewed the physics basics of spin-transfer torque magnetic tunnel junctions. We put a particular focus on the switching delay of these binary devices and the intrinsic randomness of this quantity. We saw that one can distinguish three different regimes of programming current. Depending on the fact that one uses low, intermediate, or high programming current, the reversal process is ruled by different dominant physical phenomena, which modifies the statistical distribution of the switching delay.

In a second part, we reinterpreted STT-MTJs' behavior as a stochastic memristive synapse. We studied how to achieve probabilistic programming of this kind of devices, by using the intrinsic randomness of their switching delay. We introduced the example of a spiking neural network-inspired system that can exploit this stochastic effect to perform unsupervised learning, through a simplified stochastic STDP learning rule.

The switching probabilities of the nanodevices do not need to be controlled perfectly, as the system is robust to device mismatch, which is evidenced by Monte Carlo simulations. The three programming regimes are not equivalent: the intermediate-programming-current regime minimizes the power consumption and leads to high robustness to device variations. This regime thus appears to be ideal for the use of STT-MTJs as stochastic synapses.

This system also gives insight into an original approach to use memristive nanodevices. Intrinsic unpredictability associated with nanoscale physics is not necessarily an enemy, but can be harnessed for bioinspired computing techniques.

Acknowledgements The authors would like to thank Jérôme Larroque, Nesrine Ben Romdhane, Olivier Bichler, Christian Gamrat, Weisheng Zhao, Jacques-Olivier Klein, Sylvie Galdin-Retailleau, Thibaut Devolder, Dafiné Ravelosona, Pierre Bessiere, Jacques Droulez, Alice Mizrahi, Damir

Vodenicarevic, Joseph Friedman, and Julie Grollier. Some of the works presented within this chapter were supported by the ANR COGNISPIN (ANR-13-JS03-0004-01), the FP7 ICT BAMBI (FP7-ICT-2013-C) projects, Laboratoire d'Excellence NanoSaclay (ANR-10-LABX-0035), and the CNRS/MI DEFI NANO program.

References

1. http://sourceforge.net/p/jaer/wiki/AER%20data/
2. Andre, T., Alam, S., Gogl, D., Subramanian, C., Lin, H., Meadows, W., Zhang, X., Rizzo, N., Janesky, J., Houssameddine, D., Slaughter, J.: St-mram fundamentals, challenges, and applications. In: Custom Integrated Circuits Conference (CICC), 2013 IEEE, pp. 1–8 (2013). doi:10.1109/CICC.2013.6658449
3. Apalkov, D.M., Visscher, P.B.: Spin-torque switching: Fokker-Planck rate calculation. Phys. Rev. B **72**, 180405 (2005). doi:10.1103/PhysRevB.72.180405
4. Arthur, J.V., Boahen, K.A.: Learning in silicon: timing is everything. Adv. Neural Inf. Process. Syst. **18**, 281–1185 (2006)
5. Beach, R., Min, T., Horng, C., Chen, Q., Sherman, P., Le, S., Young, S., Yang, K., Yu, H., Lu, X., Kula, W., Zhong, T., Xiao, R., Zhong, A., Liu, G., Kan, J., Yuan, J., Chen, J., Tong, R., Chien, J., Torng, T., Tang, D., Wang, P., Chen, M., Assefa, S., Qazi, M., DeBrosse, J., Gaidis, M., Kanakasabapathy, S., Lu, Y., Nowak, J., O'Sullivan, E., Maffitt, T., Sun, J., Gallagher, W.: A statistical study of magnetic tunnel junctions for high-density spin torque transfer-MRAM (STT-MRAM). In: 2008 IEEE International Electron Devices Meeting IEDM, pp. 1–4 (2008). doi:10.1109/IEDM.2008.4796679
6. Bedau, D., Liu, H., Sun, J.Z., Katine, J.A., Fullerton, E.E., Mangin, S., Kent, A.D.: Spin-transfer pulse switching: from the dynamic to the thermally activated regime. Appl. Phys. Lett. **97**(26), 262502–262502–3 (2010). doi:10.1063/1.3532960
7. Bi, G.Q., Poo, M.M.: Synaptic modification by correlated activity: Hebb's Postulate Revisited. Annu. Rev. Neurosci. **24**(1), 139–166 (2001). doi:10.1146/annurev.neuro.24.1.139
8. Bichler, O., Querlioz, D., Thorpe, S.J., Bourgoin, J.P., Gamrat, C.: Extraction of temporally correlated features from dynamic vision sensors with spike-timing-dependent plasticity. Neural Netw. **32**, 339–348 (2012). doi:10.1016/j.neunet.2012.02.022
9. Bichler, O., Suri, M., Querlioz, D., Vuillaume, D., DeSalvo, B., Gamrat, C.: Visual pattern extraction using energy-efficient "2-PCM Synapse" neuromorphic architecture. IEEE Trans. Electron Devices **59**(8), 2206–2214 (2012). doi:10.1109/TED.2012.2197951
10. Brown, W.F.: Thermal fluctuations of a single-domain particle. Phys. Rev. **130**, 1677–1686 (1963). doi:10.1103/PhysRev.130.1677
11. Chung, S., Rho, K.M., Kim, S.D., Suh, H.J., Kim, D.J., Kim, H., Lee, S., Park, J.H., Hwang, H.M., Hwang, S.M., et al.: Fully integrated 54nm stt-ram with the smallest bit cell dimension for high density memory application. In: 2010 IEEE International Electron Devices Meeting (IEDM), pp. 7–12. IEEE (2010)
12. Devolder, T., Hayakawa, J., Ito, K., Takahashi, H., Ikeda, S., Crozat, P., Zerounian, N., Kim, J.V., Chappert, C., Ohno, H.: Single-shot time-resolved measurements of nanosecond-scale spin-transfer induced switching: stochastic versus deterministic aspects. Phys. Rev. Lett. **100**(5), 057206 (2008). doi:10.1103/PhysRevLett.100.057206
13. Diao, Z., Li, Z., Wang, S., Ding, Y., Panchula, A., Chen, E., Wang, L.C., Huai, Y.: Spin-transfer torque switching in magnetic tunnel junctions and spin-transfer torque random access memory. J. Phys. Condens. Matter **19**(16), 165209 (2007). doi:10.1088/0953-8984/19/16/165209
14. Diao, Z., Li, Z., Wang, S., Ding, Y., Panchula, A., Chen, E., Wang, L.C., Huai, Y.: Spin-transfer torque switching in magnetic tunnel junctions and spin-transfer torque random access memory. J. Phys. Condens. Matter **19**(16), 165209 (2007)

15. Gaba, S., Sheridan, P., Zhou, J., Choi, S., Lu, W.: Stochastic memristive devices for computing and neuromorphic applications. Nanoscale **5**(13), 5872–5878 (2013). doi:10.1039/C3NR01176C
16. García-Palacios, J.L., Lázaro, F.J.: Langevin-dynamics study of the dynamical properties of small magnetic particles. Phys. Rev. B **58**, 14937–14958 (1998). doi:10.1103/PhysRevB.58.14937
17. Higo, Y., Yamane, K., Ohba, K., Narisawa, H., Bessho, K., Hosomi, M., Kano, H.: Thermal activation effect on spin transfer switching in magnetic tunnel junctions. Appl. Phys. Lett. **87**(8), 082502 (2005). doi:10.1063/1.2011795
18. Indiveri, G., Linares-Barranco, B., Legenstein, R., Deligeorgis, G., Prodromakis, T.: Integration of nanoscale memristor synapses in neuromorphic computing architectures. Nanotechnology **24**(38), 384010 (2013). doi:10.1088/0957-4484/24/38/384010
19. Jo, S.H., Chang, T., Ebong, I., Bhadviya, B.B., Mazumder, P., Lu, W.: Nanoscale memristor device as synapse in neuromorphic systems. Nano Lett. **10**(4), 1297–1301 (2010). doi:10.1021/nl904092h
20. Julliere, M.: Tunneling between ferromagnetic films. Phys. Lett. A **54**(3), 225–226 (1975)
21. Kavehei, O.: Highly Scalable Neuromorphic Hardware with 1-bit Stochastic nano-Synapses. arXiv e-print 1309.6419 (2013)
22. Khvalkovskiy, A., Apalkov, D., Watts, S., Chepulskii, R., Beach, R., Ong, A., Tang, X., Driskill-Smith, A., Butler, W., Visscher, P., et al.: Basic principles of stt-mram cell operation in memory arrays. J. Phys. D: Appl. Phys. **46**(7), 74001–74020 (2013)
23. Kim, J.H., Lim, W., Pi, U., Lee, J., Kim, W., Kim, J., Kim, K., Park, Y., Park, S., Kang, M., Kim, Y., Kim, W., Kim, S., Park, J., Lee, S., Lee, Y., Yoon, J., Oh, S., Park, S., Jeong, S., Nam, S., Kang, H., Jung, E.: Verification on the extreme scalability of stt-mram without loss of thermal stability below 15 nm mtj cell. In: 2014 Symposium on VLSI Technology (VLSI-Technology): Digest of Technical Papers, pp. 1–2 (2014). doi:10.1109/VLSIT.2014.6894366
24. Kitagawa, E., Fujita, S., Nomura, K., Noguchi, H., Abe, K., Ikegami, K., Daibou, T., Kato, Y., Kamata, C., Kashiwada, S., Shimomura, N., Ito, J., Yoda, H.: Impact of ultra low power and fast write operation of advanced perpendicular mtj on power reduction for high-performance mobile cpu. In: 2012 IEEE International Electron Devices Meeting (IEDM), pp. 29.4.1–29.4.4 (2012). doi:10.1109/IEDM.2012.6479129
25. Kondo, Y., Sawada, Y.: Functional abilities of a stochastic logic neural network. IEEE Trans. Neural Netw. **3**(3), 434–443 (1992). doi:10.1109/72.129416
26. Lakys, Y., Zhao, W.S., Devolder, T., Zhang, Y., Klein, J.O., Ravelosona, D., Chappert, C.: Self-enabled "Error-Free" switching circuit for spin transfer torque MRAM and logic. IEEE Trans. Magn. **48**(9), 2403–2406 (2012). doi:10.1109/TMAG.2012.2194790
27. Lecerf, G., Tomas, J., Saighi, S.: Excitatory and Inhibitory Memristive Synapses for Spiking Neural Networks. In: 2013 IEEE International Symposium on Circuits and Systems (ISCAS), pp. 1616–1619 (2013). doi:10.1109/ISCAS.2013.6572171
28. Lee, J.H., Likharev, K.K.: Defect-tolerant nanoelectronic pattern classifiers. Int. J. Circ. Theor. Appl. **35**(3), 239–264 (2007). doi:10.1002/cta.410
29. Li, Z., Zhang, S.: Thermally assisted magnetization reversal in the presence of a spin-transfer torque. Phys. Rev. B **69**, 134416 (2004). doi:10.1103/PhysRevB.69.134416
30. Lichtsteiner, P., Posch, C., Delbruck, T.: A 128x 128 120 dB 15 mus Latency asynchronous temporal contrast vision sensor. IEEE J. Solid-State Circ. **43**(2), 566–576 (2008). doi:10.1109/JSSC.2007.914337
31. Markram, H., Lubke, J., Frotscher, M., Sakmann, B.: Regulation of synaptic efficacy by coincidence of postsynaptic APs and EPSPs. Science **275**(5297), 213–215 (1997). doi:10.1126/science.275.5297.213
32. Néel, L.: Théorie du traînage magnétique des ferromagnétiques en grains fins avec applications aux terres cuites. Ann. géophys **5**(2), 99–136 (1949)
33. Nessler, B., Pfeiffer, M., Buesing, L., Maass, W.: Bayesian computation emerges in generic cortical microcircuits through spike-timing-dependent plasticity. PLoS Comput. Biol. **9**(4) (2013). doi:10.1371/journal.pcbi.1003037

34. Noguchi, H., Kushida, K., Ikegami, K., Abe, K., Kitagawa, E., Kashiwada, S., Kamata, C., Kawasumi, A., Hara, H., Fujita, S.: A 250-mhz 256b-i/o 1-mb stt-mram with advanced perpendicular mtj based dual cell for nonvolatile magnetic caches to reduce active power of processors. In: 2013 Symposium on VLSI Technology (VLSIT), pp. C108–C109 (2013)
35. Ohno, H., Endoh, T., Hanyu, T., Kasai, N., Ikeda, S.: Magnetic tunnel junction for nonvolatile cmos logic. In: 2010 IEEE International Electron Devices Meeting (IEDM), pp. 9.4.1–9.4.4 (2010). doi:10.1109/IEDM.2010.5703329
36. Querlioz, D., Bichler, O., Vincent, A., Gamrat, C.: Bioinspired programming of memory devices for implementing an inference engine. Proc. IEEE **103**(8), 1398–1416 (2015). doi:10.1109/JPROC.2015.2437616
37. Ralph, D.C., Stiles, M.D.: Spin transfer torques. ArXiv e-prints (2007)
38. Rizzo, N., Houssameddine, D., Janesky, J., Whig, R., Mancoff, F., Schneider, M., DeHerrera, M., Sun, J., Nagel, K., Deshpande, S., et al.: A fully functional 64 mb ddr3 st-mram built< newline/> on 90 nm cmos technology. IEEE Trans. Magn. **49**(7), 4441–4446 (2013)
39. Saïghi, S., Mayr, C.G., Serrano-Gotarredona, T., Schmidt, H., Lecerf, G., Tomas, J., Grollier, J., Boyn, S., Vincent, A.F., Querlioz, D., La Barbera, S., Alibart, F., Vuillaume, D., Bichler, O., Gamrat, C., Linares-Barranco, B.: Plasticity in memristive devices for spiking neural networks. Front. Neurosci **9**, 51 (2015). doi:10.3389/fnins.2015.00051
40. Senn, W., Fusi, S.: Convergence of stochastic learning in perceptrons with binary synapses. Phys. Rev. E **71**(6), 061907 (2005). doi:10.1103/PhysRevE.71.061907
41. Slonczewski, J.: Current-driven excitation of magnetic multilayers. J. Magn. Magn. Mater. **159**(1), L1–L7 (1996). doi:10.1016/0304-8853(96)00062-5
42. Snider, G.: Spike-timing-dependent learning in memristive nanodevices. In: Proceedings of IEEE International Symposium on Nanoscale Architectures 2008 (NANOARCH), pp. 85–92 (2008). doi:10.1109/NANOARCH.2008.4585796
43. Strukov, D.B., Snider, G.S., Stewart, D.R., Williams, R.S.: The missing memristor found. Nature **453**(7191), 80–83 (2008). doi:10.1038/nature06932
44. Sun, J.: Spin angular momentum transfer in current-perpendicular nanomagnetic junctions. IBM J. Res. Dev. **50**(1), 81–100 (2006). doi:10.1147/rd.501.0081
45. Sun, J.Z.: Spin-current interaction with a monodomain magnetic body: a model study. Phys. Rev. B **62**, 570–578 (2000). doi:10.1103/PhysRevB.62.570
46. Sun, J.Z., Robertazzi, R.P., Nowak, J., Trouilloud, P.L., Hu, G., Abraham, D.W., Gaidis, M.C., Brown, S.L., O'Sullivan, E.J., Gallagher, W.J., Worledge, D.C.: Effect of subvolume excitation and spin-torque efficiency on magnetic switching. Phys. Rev. B **84**, 064413 (2011). doi:10.1103/PhysRevB.84.064413
47. Suri, M., Bichler, O., Querlioz, D., Cueto, O., Perniola, L., Sousa, V., Vuillaume, D., Gamrat, C., DeSalvo, B.: Phase change memory as synapse for ultra-dense neuromorphic systems: application to complex visual pattern extraction. In: IEDM Technical Digest, pp. 4.4.1–4.4.4. IEEE (2011). doi:10.1109/IEDM.2011.6131488
48. Suri, M., Bichler, O., Querlioz, D., Palma, G., Vianello, E., Vuillaume, D., Gamrat, C., DeSalvo, B.: CBRAM devices as binary synapses for low-power stochastic neuromorphic systems: auditory (Cochlea) and visual (Retina) cognitive processing applications. IEDM Technical Digest, p. 10.3.1 (2012)
49. Suri, M., Querlioz, D., Bichler, O., Palma, G., Vianello, E., Vuillaume, D., Gamrat, C., DeSalvo, B.: Bio-Inspired stochastic computing using binary CBRAM synapses. IEEE Trans. Electron Devices **60**(7), 2402–2409 (2013). doi:10.1109/TED.2013.2263000
50. Thomas, L., Jan, G., Zhu, J., Liu, H., Lee, Y.J., Le, S., Tong, R.Y., Pi, K., Wang, Y.J., Shen, D., He, R., Haq, J., Teng, J., Lam, V., Huang, K., Zhong, T., Torng, T., Wang, P.K.: Perpendicular spin transfer torque magnetic random access memories with high spin torque efficiency and thermal stability for embedded applications (invited). J. Appl. Phys. **115**(17), 172615 (2014). doi:10.1063/1.4870917
51. Vincent, A., Larroque, J., Locatelli, N., Ben Romdhane, N., Bichler, O., Gamrat, C., Zhao, W., Galdin-Retailleau, S., Querlioz, D.: Spin-transfer torque magnetic memory as a stochastic memristive synapse for neuromorphic systems. IEEE Trans. Biomed. Circuits Syst. **7**(2), 166–174 (2015). doi:10.1109/TBCAS.2015.2414423

52. Vincent, A., Locatelli, N., Klein, J.O., Zhao, W., Galdin-Retailleau, S., Querlioz, D.: Analytical macrospin modeling of the stochastic switching time of Spin-Transfer Torque devices. IEEE Trans. Electron Devices **62**(1), 164–170 (2015). doi:10.1109/TED.2014.2372475
53. Wolf, S.A., Lu, J., Stan, M.R., Chen, E., Treger, D.M.: The promise of nanomagnetics and spintronics for future logic and universal memory. Proc. IEEE **98**(12), 2155–2168 (2010). doi:10.1109/JPROC.2010.2064150
54. Worledge, D., Hu, G., Trouilloud, P., Abraham, D., Brown, S., Gaidis, M., Nowak, J., O'Sullivan, E., Robertazzi, R., Sun, J., Gallagher, W.: Switching distributions and write reliability of perpendicular spin torque MRAM. In: 2010 IEEE International Electron Devices Meeting (IEDM), pp. 12.5.1–12.5.4 (2010). doi:10.1109/IEDM.2010.5703349
55. Zhang, Y., Zhao, W., Prenat, G., Devolder, T., Klein, J.O., Chappert, C., Dieny, B., Ravelosona, D.: Electrical modeling of stochastic spin transfer Torque writing in magnetic tunnel junctions for memory and logic applications. IEEE Trans. Magn. **49**(7), 4375–4378 (2013). doi:10.1109/TMAG.2013.2242257
56. Zhao, W., Chappert, C., Javerliac, V., Noziere, J.P.: High speed, high stability and low power sensing amplifier for mtj/cmos hybrid logic circuits. IEEE Trans. Magn. **45**(10), 3784–3787 (2009). doi:10.1109/TMAG.2009.2024325

Multiple Binary OxRAMs as Synapses for Convolutional Neural Networks

E. Vianello, D. Garbin, O. Bichler, G. Piccolboni, G. Molas, B. De Salvo and L. Perniola

Abstract Oxide-based resistive memory (OxRAM) devices find applications in memory, logic, and neuromorphic computing systems. Among the different dielectrics proposed in OxRAM stacks, hafnium oxide, HfO_2, attracted growing interest because of its compatibility with typical BEOL advanced CMOS processing and promising performances in terms of endurance (higher than Flash) and switching speed (few tens of ns). This chapter describes an artificial synapse composed of multiple binary HfO_2-based OxRAM cells connected in parallel, thereby providing synaptic analog behavior. The VRRAM technology is presented as a possible solution to gain area with respect to planar approaches by realizing one VRRAM pillar per synapse. The HfO_2-based OxRAM synapse has been proposed for hardware implementation of power efficient Convolutional Neural Networks for visual pattern recognition applications. Finally, the synaptic weight resolution and the robustness to device variability of the network have been investigated. Statistical evaluation of device variability is obtained on a 16 kbit OxRAM memory array integrated into advanced 28 nm CMOS technology.

1 Multiple Binary OxRAM Devices as Artificial Synapses

Research activities in the field of brain-inspired computation have gained importance in recent years [1–4]. Emerging backend resistive memory devices are considered the optimum candidates to emulate biological synaptic behavior at nanometer scale, thanks to the fact that they offer the possibility to modulate their conductance by applying low biases, and they can be easily integrated with CMOS-based neuron circuits [5]. This opens the way to the realization of compact and energy-efficient computing architectures based on artificial neural networks. In literature, several

E. Vianello (✉) · D. Garbin · G. Piccolboni · G. Molas · B. De Salvo · L. Perniola
CEA LETI MINATEC Campus, 17 Rue des Martyrs, 38054 Grenoble Cedex 9, France
e-mail: elisa.vianello@cea.fr

O. Bichler
CEA LIST, 91191 Gif-sur-yvette, France

© Springer (India) Pvt. Ltd. 2017

M. Suri (ed.), *Advances in Neuromorphic Hardware Exploiting Emerging Nanoscale Devices*, Cognitive Systems Monographs 31,
DOI 10.1007/978-81-322-3703-7_6

Resistive RAM (RRAM) technologies such as phase-change memory (PCM), conductive-bridge memory (CBRAM), and oxide-based resistive memory (OxRAM) have been investigated as possible solutions for the implementation of artificial synapses [6–10]. We focused on HfO_2-based OxRAM technology [11, 12], which demonstrated attractive features such as low switching voltage and fast switching speed (few tens of ns at 1 V), promising endurance (up to 10^8 cycles) and high scalability ($10 \cdot 10$ nm^2) [13].

Two main approaches to emulate synaptic conductance modulation were successfully demonstrated using RRAM devices:

1. *analog approach*, where multiple low-resistance states for emulating long-term potentiation (cumulative increase of conductance, LTP) and multiple high-resistance states for long-term depression (cumulative and gradual decrease of conductance, LTD) were adopted [7, 8];
2. *binary approaches*, where only two distinct resistive states (low-resistance state, LRS and high-resistance state, HRS) per device associated with a probabilistic STDP bioinspired learning rule were adopted [6, 14].

HfO_2/Ti OxRAM cells are intrinsically binary devices: They switch between two distinct resistance states, a low-resistance state and a high-resistance state when appropriate identical SET/RESET pulses are applied. These programming pulses can be optimized for high speed and low power consumption, and are thus ideal for the implementation of an energy-efficient hardware implementation of artificial neural networks. To achieve multiple resistance levels adopting HfO_2/Ti OxRAM cells, each neuron must generate pulses with increasing amplitude to gradually increase the programming current [15]. This implies keeping a history of the previous state of the synaptic device, thus leading to additional overhead in the programming circuitry. For this reason, we adopted the binary approach. However, the use of only two resistance levels per synapse, with respect to the multilevel approach, can be insufficient to achieve good performances in neuromorphic systems designed for some complex applications, as for example image recognition [16]. Consequently, to emulate synaptic conductance, we propose a solution based on a *hybrid approach*, which tries to unify the advantages of both multilevel and binary approaches. In this solution, a single synapse is composed of n multiple binary OxRAM cells operating in parallel ($nTnR$ structure) [15]. The model which we refer to is schematically represented in Fig. 1; all the devices on the same row, connected in parallel, build an equivalent synapse which connects a presynaptic neuron (neuron A) to a post-synaptic neuron (neuron B). Since parallel conductance sum-up, the conductance of the equivalent synapse ranges from the sum of the n conductances in the HRS to the sum of all the n conductances in the LRS. This strategy provides the opportunity to build an analog-like conductance behavior for a binary device, at the cost of an increased number of devices needed to build a synapse. This approach offers the advantage of a simple programming methodology for the OxRAM devices, in which standard SET and RESET pulses, optimized for high endurance and low power consumption, are used to switch the device resistance from LRS to HRS and vice versa.

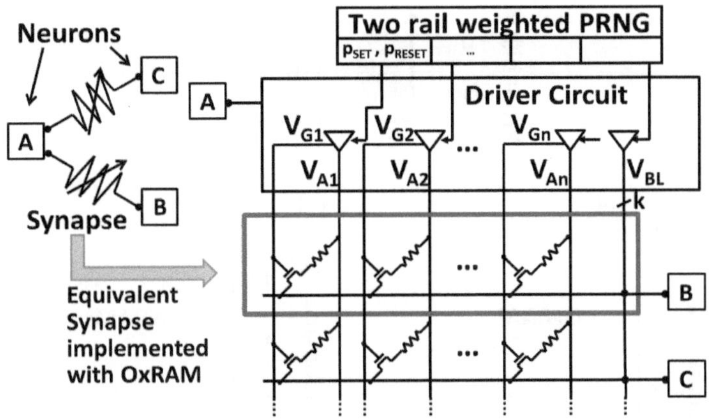

Fig. 1 Schematic of OxRAM-based synapse. All the OxRAM devices on the same row build one equivalent synapse ($nTnR$ structure). Driver circuit is used to individually program OxRAM devices and propagate spikes to next neuron layer. The weighted PRNG is used for online learning, to implement extrinsic stochasticity in probabilistic STDP learning rule. (@ 2016 IEEE. Reprinted with permission from [15])

In order to define the resistance state (LRS or HRS) of each OxRAM device needed to obtain the desired equivalent synaptic conductance, two alternative approaches can be used: supervised or unsupervised learning. *Supervised learning* is obtained using back-propagation algorithm [17], where the LRS/HRS status of each OxRAM device is determined with computer simulations (offline learning), and then discretized and imported in the memory array with a one-time programming operation. In *unsupervised learning*, the LRS/HRS status of the devices is learned in-situ (online learning) with the stochastic STDP learning rule shown in Fig. 2a [18]. According to the difference Δt of the spiking time of the post-neuron (t_s) and the preneuron (t_x), a Long-Term Potentiation (LTP) or a Long-Term Depression (LTD) operation is carried out. An LTP (LTD) operation consists in applying to each device of the equivalent synapse a SET (RESET) operation with a probability p_{SET} (p_{RESET}). The switching probability can be governed by the RRAM itself (internal switching probability): SET and RESET conditions can be tuned to control the probability to switch the memory [6]. Another possibility, which allows a more fine-tuning of the switching probability, at the expense of small increase of the circuit complexity, consists in using stronger programming conditions that do not show intrinsic stochasticity; i.e., switching probability is equal to one. Extrinsic stochasticity is thus obtained using an external Pseudorandom Number Generator (PRNG) circuit block, which provides tunable switching probabilities p_{SET} and p_{RESET}. The driver circuit block can be used to individually program the OxRAM devices (see Fig. 1).

In order to validate the functionality of the proposed synapse design (Fig. 1), we carried out simulations of LTP and LTD operations on OxRAM synapses composed by a variable number of devices connected in parallel. Figure 2b–e show the evolution of the conductance corresponding to 100 LTP (SET) followed by 100 LTD (RESET)

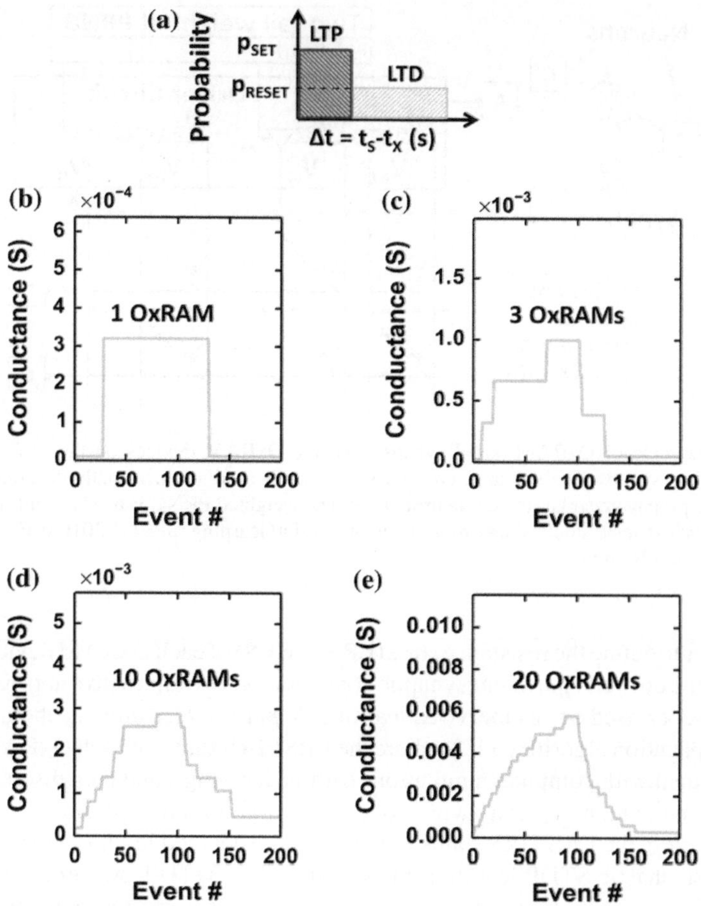

Fig. 2 **a** Probabilistic STDP learning rule. 100 consecutive LTP (SET) and LTD (RESET) events, with $p_{SET} = 0.02$ and $p_{RESET} = 0.04$ on a synapse composed of **b** 1 OxRAM device, **c** 3 OxRAM devices, **d** 10 OxRAM devices, and **e** 20 OxRAM devices connected in parallel, as shown in Fig. 1. The use of multiple devices allows to implement a multilevel equivalent synapse, and increasing the number of devices connected in parallel increases the number of intermediate conductance levels. It should be noted that the vertical axis scale is not constant. (@ 2016 IEEE. Reprinted with permission from [15])

operations for a synapse composed of $n = 1$ OxRAM device, $n = 3$ OxRAM devices, $n = 10$ OxRAM devices, and $n = 20$ OxRAM devices connected in parallel, using a binary probabilistic approach with $p_{SET} = 0.02$ and $p_{RESET} = 0.04$. The LTP (SET) and LTD (RESET) programming conditions are targeted to obtain 3 and 30 kΩ LRS and HRS, respectively. In the case of Fig. 2b, a single OxRAM device, obviously, only two conductance levels can be achieved. Using multiple OxRAM devices, Fig. 2c–e, allows to obtain a gradual modulation of conductance, with a behavior that is similar to an analog approach. Increasing the number of devices connected in parallel increases the number of intermediate conductance levels. For the same number of intermediate conductance levels, using multiple OxRAM devices does not necessarily introduce a penalty in power consumption with respect to a single analog device. The number of switching events needed to program the synaptic weight is the same in the case of single analog synapse and multiple binary OxRAMs (it is *n* switching events times 1 device and 1 switching event times *n* devices for the analog and binary approaches, respectively). Achieving multiple conductance levels with multiple devices in parallel has the advantage of enabling a multilevel behavior in a way which is independent on technology: The first target for RRAM memories for Flash replacement is to achieve two distinct resistance levels.

To study the impact of the programming conditions (power consumption) on the synaptic behavior, we calculated the cumulative distributions of LRS and HRS corresponding to weak and strong programming conditions (Fig. 3a, b). Weak programming conditions (switching energy/device ~9 pJ) result in smaller programming window (i.e., smaller separation between the distributions of HRS and LRS)

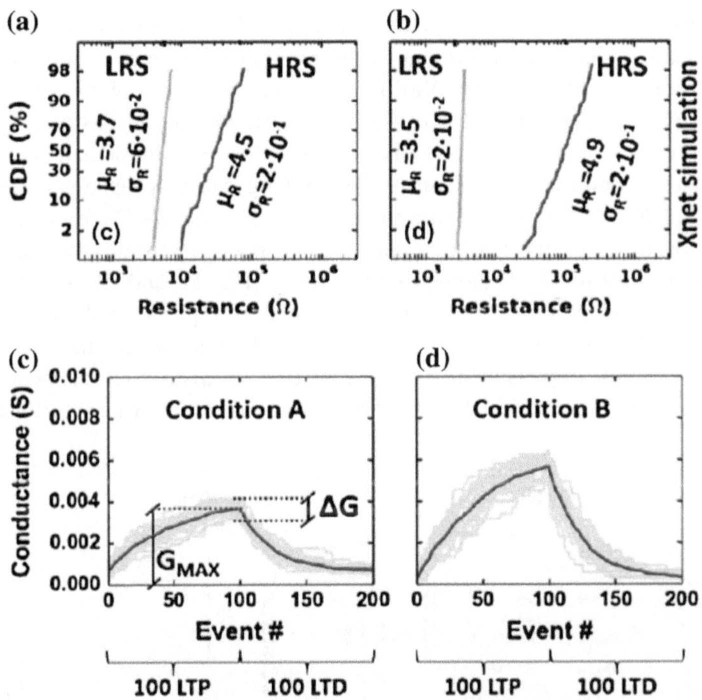

Fig. 3 Experimental resistance levels and associated variability for **a** weak ($I_{comp} = 50\,\mu A$, $V_{set} = 1V$, $V_{reset} = 1.3\,V$, $T_{pulse} = 100$ ns,) and **b** strong ($I_{comp} = 340\,\mu A$, $V_{set} = 1$ V, $V_{reset} = 1.7$ V, $T_{pulse} = 100$ ns) programming conditions. Conductance evolution corresponding to 100 consecutive LTP (SET) and LTD (RESET) events. *Gray lines* are representative of 25 synapses composed of 20 OxRAMs each, programmed with **a** weak programming conditions and **b** strong programming conditions (@ 2016 IEEE. Reprinted with permission from [15])

and larger variability. Stronger programming conditions (switching energy/device ~58 pJ), on the other hand, result in larger programming window and tighter distributions showing better variability for LRS. Figure 3c, d shows the impact of the choice of the programming condition on the conductance evolution of the synapses. Light gray curves are the conductance response of 25 synapses composed of 20 OxRAM devices each, when 100 LTP and 100 LTD operations are performed consecutively with $p_{SET} = 0.02$ and $p_{RESET} = 0.04$. Red curves are the mean conductance over 25 synapses. When stronger programming conditions are used (condition B), the associated larger programming window allows achieving a wider range of conductance values with respect to weaker programming conditions (condition A). The quantities G_{MAX}, i.e., the average conductance after 100 LTP events, and ΔG, i.e., the difference between the maximum and minimum conductances on a set of 25 synapses, have been extracted for the two conditions. Due to the fact that a probabilistic learning rule is used, the impact of the device variability on the synaptic conductance response plays a secondary role with respect to the stochasticity introduced by the probabilistic STDP learning rule. In fact, a ratio $\Delta G/G_{max} \sim 32\,\%$ is obtained for both programming conditions.

The weakest point of the proposed *hybrid approach* is the silicon area consumption for each synapse that is proportional to the number of devices needed to obtain a multilevel behavior. A possible solution to overcome this problem is the adoption of the Vertical RRAM (VRRAM) technology [19], which consists of RRAM cells integrated in multilayered VNAND-like structure, this is a simple and cost-effective 3D processes to achieve high memory density (Fig. 4). A synapse is composed by n stacked VRRAM with one common select transistor, $1TnR$ structure (Fig. 4b). This solution offers significant area gain with respect to neural networks in planar configuration with n $1T1R$ elements in parallel (Fig. 1), and the silicon area for each synapse is independent of the number of devices required to obtain a multilevel behavior. The $1TnR$ structure imposes the use of the RRAM intrinsic variability in order to implement progressive on line learning (the use of an external Pseudorandom Number Generator circuit block is possible only with a $nTnR$ structure). Figure 4c

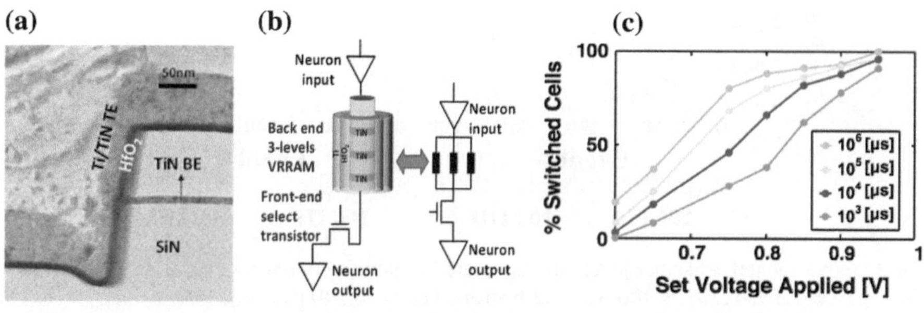

Fig. 4 **a** TEM cross sections on one TiN/HfO_2/Ti/TiN VReRAM. **b** $1TnR$ VReRAM pilar for one synapse. **c** Percentage of SET cells as a function of the pulse voltage for different pulse times (@ 2016 IEEE. Reprinted with permission from [19])

shows the percentage of switched cells (~50 VRRAM measured) as a function of the applied bias and pulse times [19]. Programming conditions can be identified to control the probability to switch the memory with a given value for each pulse, the probability being imposed by the application and neural network structure.

2 Convolutional Neural Network Architecture

The implementation of artificial neural networks composed of CMOS neurons, and NVM-based synapses have been widely investigated in the literature [3, 4, 21, 22]. The network topology that has been mostly investigated is the *fully connected neural network*. In this topology, neurons are organized in layers. Each neuron of the N layer is connected to every neuron of the $N+1$ layer by a large number of synapses (Fig. 5a). The first neuron layer is connected to the input of the network, while the last neuron layer represents the output of the system. The neuron layers between input and output are generally referred to as *hidden* layers. Fully connected neural network topologies are often limited to a maximum number of hidden layers equal to one or two. Further increasing the number of layers explosively increases the complexity of the network and the number of required synapses, without necessarily improving the performance of the network. *Convolutional neural networks* (CNNs), often referred to as *deep neural networks*, are composed of a cascade of many layers. The first layers of a CNN are convolutional layers, with a topology schematized in Fig. 5b. Neurons of a convolutional layer are organized in feature maps. Neurons in one feature map receive inputs from a small subset of neurons (receptive field) in the previous layer and produce an output which is a threshold or sigmoidal function of the weighted sum of inputs. The connectivity pattern between the neurons of the receptive field of one layer and the neurons of the subsequent layer, responsible for the weighted sum operation, forms the convolution kernel. The latter is composed of a small set of synapses shared among different neurons to connect layers N and $N+1$ through a

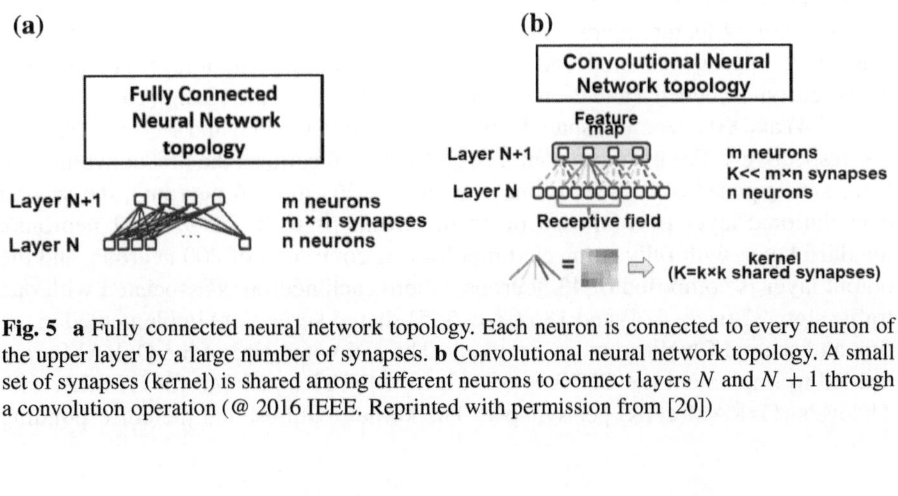

Fig. 5 **a** Fully connected neural network topology. Each neuron is connected to every neuron of the upper layer by a large number of synapses. **b** Convolutional neural network topology. A small set of synapses (kernel) is shared among different neurons to connect layers N and $N+1$ through a convolution operation (@ 2016 IEEE. Reprinted with permission from [20])

convolution operation. The kernel corresponds to a feature that has to be localized in the input image. A peak in the convolution signal means that the feature is present in the input pattern, and the feature map indicates where the feature is present in the input field. At each convolutional layer, the input pattern undergoes a transformation to a higher, more abstract representation. In the case of image recognition applications, for example, the kernel features in the first convolutional layer typically represent simple edges or segments with a given orientation, while the features of the second layer represent particular arrangements of edges in more complex shapes. After the convolutional layers, a classifier with fully connected topology is used to classify objects as combinations of the different parts extracted by the previous convolutional layers. The organization of convolutional layers in CNNs is originally inspired by the structure of the visual system in mammals [23–25]. Software implementations of CNNs were applied with great success in applications such as traffic sign recognition [26], the analysis of biological images [27], and the detection of faces, complex text, pedestrians on the streets and human bodies in natural images [28–31]. A major recent practical success of software implementations of CNNs is the face recognition software proposed by Facebook [32].

The power consumption to perform convolution operations is computationally expensive in the CNN implementation on CPUs and GPUs. This hinders their integration in portable devices. In recent years, dedicated system on chip (SOC) solutions and FPGA platforms have been used to implement these networks for increasing performances while decreasing their power consumption. A hardware implementation of CNNs based on the OxRAM devices can further improve the power efficiency. We proposed the use of OxRAM synapses presented in Sect. 1 to store the kernel features [20, 33]. Two different visual pattern recognition applications were demonstrated: the MNIST handwritten digits database [34] and the German Traffic Sign Recognition Benchmark (GTSRB) database [35]. For both the applications, the proposed architecture is composed of a feature extraction module, made of two cascaded convolutional layers, each of them followed by a subsampling layer [36] in the case of the GTSRB network, and a classification module, made of two fully connected layers (Fig. 6). For the MNIST applications, the first convolutional layer is composed of 16 feature maps of size $13 \cdot 13$ (169 neurons), the second convolutional layer is composed of 24 feature maps with size $5 \cdot 5$ (25 neurons), the third layer, with fully connected topology, is composed of 150 neurons, and the output layer is composed of 10 neurons, where each neuron is associated with one of the 10 digit categories. 16 (size: 4·4) and 90 (size: 5·5) shared kernels are used in the first and second CNN layers, respectively. For the more complex GTSRB applications, the first convolutional layer is composed of 32 feature maps with size $26 \cdot 26$ (676 neurons), the second convolutional layer is composed of 48 feature maps with size $9 \cdot 9$ (81 neurons), the third layer, with fully connected topology, is composed of 200 neurons, and the output layer is composed of 43 neurons, where each neuron is associated with one traffic sign. 32 (size: $4 \cdot 4$) and 186 (size: $5 \cdot 5$) shared kernels are implemented in the first and second CNN layers, respectively. The estimated size of the OxRAM array needed to implement the CNNs is 600 kb for MNIST and 1 Mb for GTSRB, with 11 levels (OxRAM cells) per synapse. The kernel features, and the corresponding

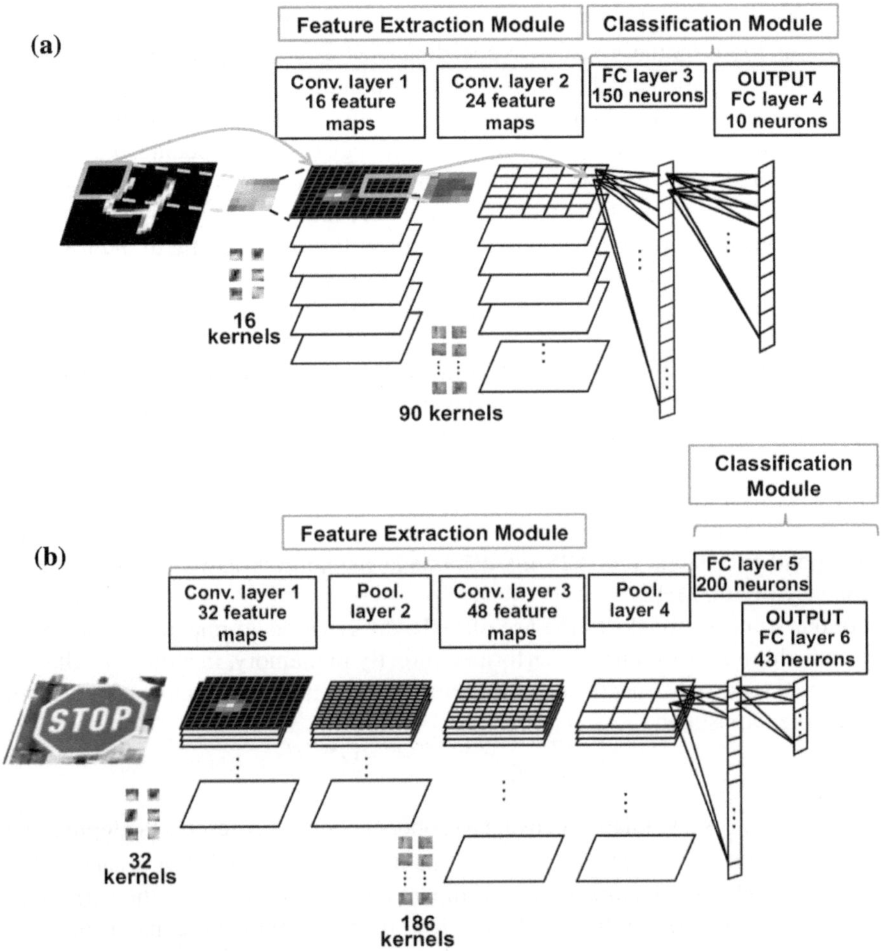

Fig. 6 CNN architecture for **a** handwritten digits recognition (MNIST database) and **b** traffic signs recognition (GTSRB database). (@ 2016 IEEE. Reprinted with permission from [33])

synaptic weights, must be learned in an initial phase and then the network can be used in read mode for visual pattern recognition. The synaptic weights are defined *offline* with supervised back-propagation learning algorithm [17].

The designed architecture has a structure equivalent to the one used for software implementations of CNNs. The main difference resides in the way the convolution operation is carried out. Mathematically, a discrete convolution operation, for a kernel with size $k \cdot k$, is described by the following equation:

$$O_{i,j} = \sum_{p=0}^{k-1} \sum_{q=0}^{k-1} I_{i+q,j+p} \cdot K_{p,q} \tag{1}$$

where $O_{i,j}$ is the brightness value at coordinates (i, j) in the output feature map, $I_{i+q,j+p}$ is the brightness values at coordinates $(i+p, j+q)$ in the input feature map, and $K_{p,q}$ is the kernel coefficient at coordinates (p, q).

In a conventional Von Neumann architecture, the convolution operation would be carried out using digital multipliers, adders, and registers. The operands $I_{i+q,j+p}$ and $K_{p,q}$ are stored in memory as numbers in digital format. At every clock cycle, these data have to be retrieved from the system memory and stored back in memory after computation. This process has to be repeated N_c times, according to the following equation:

$$N_c = k^2 \cdot f^2 \cdot N_k \cdot N_F \cdot N_{cl} \tag{2}$$

where k is the size of one kernel, f is the size of the input feature map, N_k is the number of kernels in one convolutional layer, N_F is the number of feature maps in one convolutional layer, and N_{cl} is the number of convolutional layers in the network. In the case of a state-of-the-art convolutional neural network for the recognition of traffic signs with 8-bit synapses [37], for example, a Von Neumann implementation would require ~125 million clock cycles for the recognition of one image. This would correspond to a latency of 625 ms per image recognition assuming an operating frequency of 200 MHz.

In our solution, thanks to the use of OxRAM synapses to implement the kernel, the convolution operations are performed directly in memory, in a fully parallel and distributed approach. Specifically, the multiplications are carried out in parallel using the simple Ohms law:

$$I^{output}_{i+q,j+p} = V^{input}_{i+q,j+p} \cdot G^{kernel}_{q,p} \tag{3}$$

where $V^{input}_{i+q,j+p}$ is a voltage that, using a proper encoding, represents the input image at the pixel $(i+q, p+j)$, $G^{kernel}_{q,p}$ is the conductance of an OxRAM synapse which represents the kernel feature at the coordinate (p, q) and $I_{i+q,j+p}$ is the current that has to be accumulated at the coordinate (i, j) of the output feature map neuron.

Figure 7 shows the voltage encoding of the input image ($V^{input}_{i+q,j+p}$, in Eq. (3)) into Address Event Representation (AER) format and the propagation of spiking activity through neuron layers of the CNN architecture for the handwritten digits recognition (Fig. 6). The input images are composed of $29 \cdot 29$ pixels. Each pixel's brightness, $V^{input}_{i+q,j+p}$, is converted into a voltage spike train with a given frequency, during a time slot $t = 1\,\mu$s. The lowest pixel brightness (i.e., black pixel) is converted to the lowest spiking frequency $f_{MIN} = 1$ MHz. The highest pixel brightness (i.e., white pixel) is converted to the highest spiking frequency $f_{MAX} = 8$ MHz. All the grayscale, intermediate pixel brightness values are linearly converted into spiking frequency between f_{MIN} and f_{MAX}. Moreover each pixel is associated with a neuron address (a sequential address from 0 to 840, corresponding to the $29 \cdot 29$ pixels).

Figure 8 represents the hardware implementation of a convolutional kernel ($G^{kernel}_{q,p}$, in Eq. (3)) with OxRAM array. The kernel is a collection of $k \cdot k$ synaptic weights, representing a feature to be convoluted with the input image. Each row represents one of the $k \cdot k$ synaptic weights of the kernel, and, at each row, an

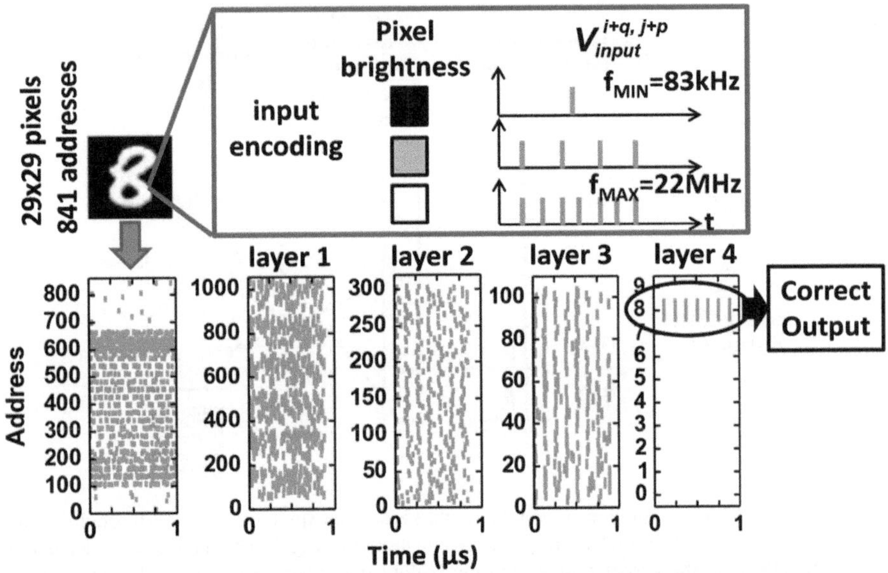

Fig. 7 Encoding of the input image ($V_{i+q,j+p}^{input}$) in the Address Event Representation (AER) format and propagation of spiking activity through neuron layers of the CNN architecture for the handwritten digits recognition (Fig. 6). (@ 2016 IEEE. Reprinted with permission from [20])

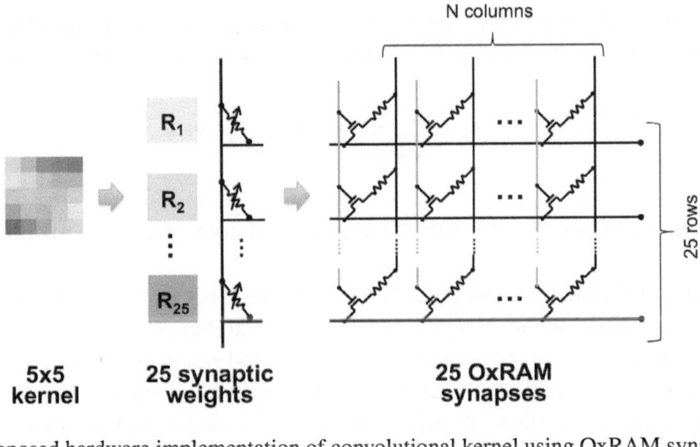

Fig. 8 Proposed hardware implementation of convolutional kernel using OxRAM synapses

OxRAM-based synapse composed of n devices connected in parallel is implemented (as illustrated in Fig. 1 and Sect. 1). Figure 9 illustrates the read-mode operation in the convolutional kernel. When a spike, $V_{i+q,j+p}^{input}$, occurs at coordinates $(i+q, j+p)$ in the input image, an address decoder is used to dynamically map the kernel synapses to the feature map neurons that have the input neuron $(i+q, j+p)$ in their receptive field, p and q ranging from 1 to $k-1$ (k is the size of the kernel). The spike is

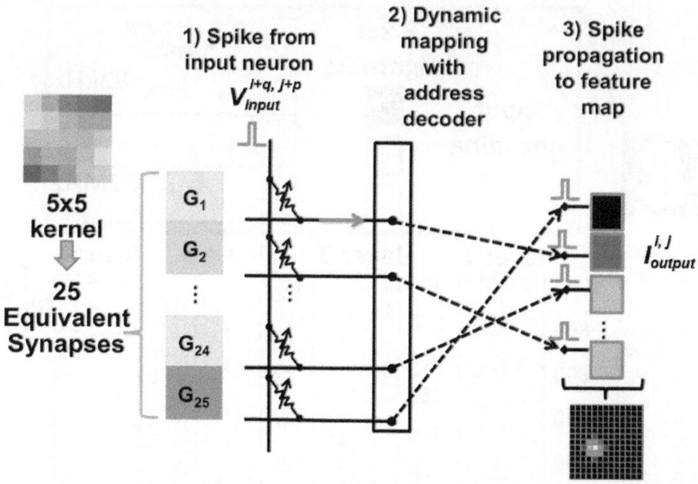

Fig. 9 Proposed implementation of convolutional kernel and spike propagation through it. Each synapse is composed of 20 OxRAM devices as illustrated in Fig. 1. The address decoder is used to dynamically map the kernel synapses to the feature map neurons that have the input neuron $(i + q, j + p)$ in their receptive field. (@ 2016 IEEE. Reprinted with permission from [20])

then propagated through the synapses of the kernel to the mapped output Integrate and Fire (IF) neurons. The IF neurons accumulate (integrate) the incoming current over time and will fire when a given threshold is reached. The accumulated current at the coordinate (i, j) of the output feature map neuron is given by the following formula:

$$I_{i,j}^{output} = \sum_{p=0}^{k-1} \sum_{q=0}^{k-1} V_{i+q,j+p}^{input} \cdot G_{p,q}^{kernel} \tag{4}$$

The spiking frequency of the feature map neuron at coordinate (i, j) is proportional to $I_{i,j}^{output}$.

In order to validate the functionality of the two CNNs architectures proposed for the recognition task of the MNIST and GTSRB databases (Fig. 6), we performed simulations using the special-purpose spiking neural network simulator Xnet [38], using synapses composed of $n = 20$ OxRAM devices connected in parallel. In order to define the resistance state of each OxRAM device, we used the supervised back-propagation learning algorithm. Figure 7b shows an example of the propagation of the spikes through the layers of the CNN of the CNN architecture for the handwritten digits recognition (Fig. 6), when a test image representing the handwritten digit 8 is presented to the network. At the input layer, the static image is converted in AER format, with neurons spiking at different frequencies according to the brightness of the corresponding image pixel. The signals are propagated through the network until the output layer, where the neuron with the highest spiking frequency (neuron number 8 in this specific case) indicates the category in which the input image has been categorized by the network.

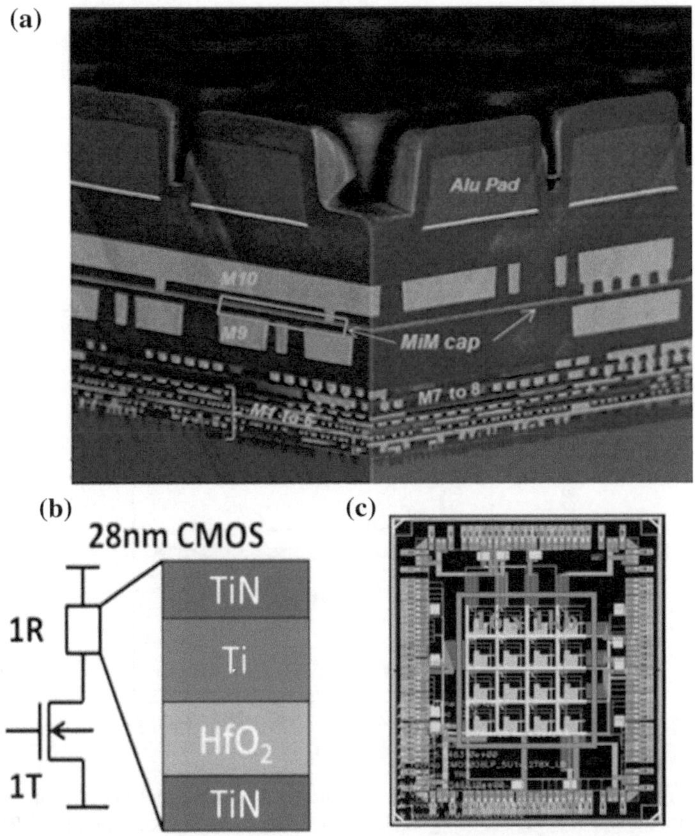

Fig. 10 **a** SEM cross section of CMOS 28 nm stack including OxRAM device, **b** 1T1R bitcell schematic, **c** 16 kbit demonstrator. (@ 2016 IEEE. Reprinted with permission from [33])

3 Synaptic Weight Resolution and Tolerance to Variability

In order to study the impact of the OxRAM electrical performances and reliability on the network, we fully characterized a 16 kbit OxRAM demonstrator integrated into a 28-nm CMOS digital test-chip (Fig. 10) [12]. OxRAM devices feature a metal–insulator–metal (MIM) structure composed of a 5-nm-thick HfO_2 layer sandwiched between a Ti top electrode and a TiN bottom electrode. A bitcell is composed of 1 Transistor–1 Resistor (1T1R) structure. The access transistor is used to select and limit the current flowing through the device during programming. Figure 11 reports the cumulative distributions of low-resistance state (LRS) and high-resistance state (HRS) extracted from the 16 kb OxRAM array statistics. No correction code or smart programming algorithms have been used. Discrete steps in the experimental distributions are given by discrete thresholds in read current sensing. The experimental distribution is cut at about 1 MΩ because of the lower limit in current sensing. All

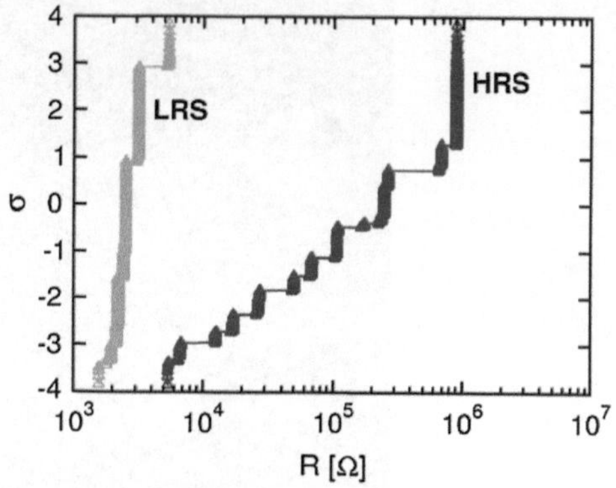

Fig. 11 Cumulative distributions of LRS and HRS for 16kb array. Discrete steps are due to discrete thresholds in read current sensing. The experimental distribution is cut at 1M due to lower limit in current sensing

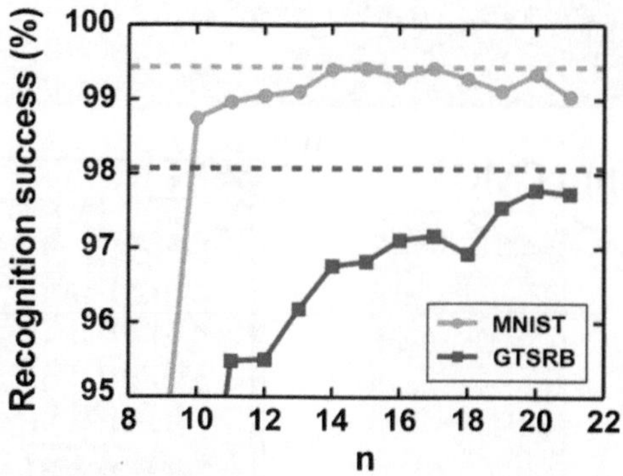

Fig. 12 Recognition success as a function of the number n of parallel OxRAM devices used to implement an equivalent synapse, using analog neuron model and taking into the LRS and HRS OxRAM distributions presented in Fig. 11. *Dashed lines* reference recognition success rate obtained on the testing dataset with the formal CNN model and floating-point precision synapses. (@ 2016 IEEE. Reprinted with permission from [33])

the network simulations presented into the following take into account the real LRS and HRS distributions presented in Fig. 11.

In Sect. 1, we demonstrated that using more OxRAM cells per synapse increases the synaptic weight resolution (Fig. 2), but comes at the cost of larger area consumption (for the $nTnR$ structure, Fig. 1) or more complex process integration ($1TnR$ VReRAM architecture, Fig. 4). We therefore studied the impact of the number, n, of OxRAM devices per synapse on the performance of the CNN architectures for the recognition tasks of the MNIST and GTSRB databases presented in Fig. 11. Parametric simulations have been performed, varying the number of OxRAM cells per synapse, n, and keeping all the other parameters of the network constant. The red curve in Fig. 12 reports the recognition success of 10 000 handwritten digits after learning with back-propagation algorithm [17] as a function of the number n. The recognition success improves as n increases, for n higher than 12 the maximum network performance greater than 97 % is reached. The blue curve in Fig. 12 reports the recognition success as a function of the number n of OxRAM devices per synapse for the GTSRB database recognition. More complex application tasks, such as the GTSRB database recognition, are more demanding in terms of synaptic weight resolution. In the case of handwritten digits recognition (MNIST), 11 OxRAM devices per synapse are enough to achieve a recognition performance equivalent to the reference recognition success rate obtained with the formal CNN model with floating-point precision for the synapses. In the case of the more complex recognition task of traffic signs, 20 OxRAM devices per synapse are necessary to achieve a recognition rate equivalent to the reference one.

OxRAMs are considered a promising technology for Flash replacement in both stand-alone and embedded memory solutions [39, 40]. However, one of the main issues for conventional memory applications is the noise behavior of the high-resistance state (HRS) that reduces the memory operation window [12]. In the following, the impact of resistance variability on the performance of Convolutional Neural Network (CNN) systems will be investigated. In order to quantify the impact of the OxRAM resistance variability on the performance of the CNN, we used the MNIST database as test bench simulations. Since the OxRAM LRS and HRS resistance distributions can be fitted by lognormal distributions with mean value μ and standard deviation σ, the resistance of each OxRAM device (R_i) in the Xnet simulations of the network is defined using the following relation:

$$R_i = \text{lognorm}(\mu, \sigma) \tag{5}$$

where lognorm (μ, σ) is a function that draws a random sample from the base-10 lognormal distribution with parameters μ and σ. The experimental HRS and LRS distribution of Fig. 11 are well reproduced by lognormal distributions with $\mu = 5.5$, $\sigma = 0.45$ and $\mu = 3.45$, $\sigma = 0.06$, respectively. These values allow to achieve a recognition success larger than 98.9 %. To study the impact of the HRS variability on the network performances, we performed Xnet simulations of the CNN network with different values for σ and μ of the HRS distribution (Figs. 13a and 14b). Figure 13b demonstrates that a reduction in the HRS variability (corresponding to a reduction of the standard deviation of the HRS distribution) does not improve the recognition rate of the network. Moreover, we studied the impact of the reduction of the HRS mean value for a given value of σ on the network performances (Fig. 14). A reduction of the HRS mean value of about one order of magnitude (μ form 5.5 to 4.5) does not compromise the recognition rate, which remains higher than 98.8 %. A further reduction of the HRS ($\mu = 3.9$) starts to degrade the network performances. These results confirm the robustness and tolerance of the proposed network to the HRS variability [41].

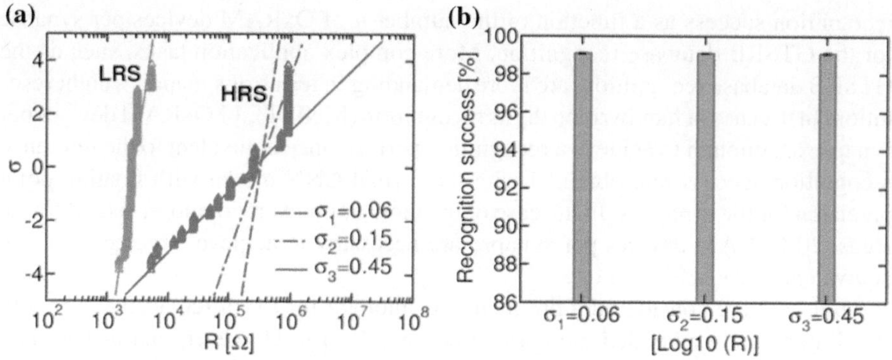

Fig. 13 **a** Distributions of LRS (*red*) and HRS (*blue*) for experimental 16 kbit (*symbols*) and HRS lognormal distributions implemented in our Xnet (*lines*) with different variabilities (σ) and same mean value (μ). **b** Impact of the standard deviation of the lognormal HRS distribution (σ) on recognition success of OxRAM-based CNN

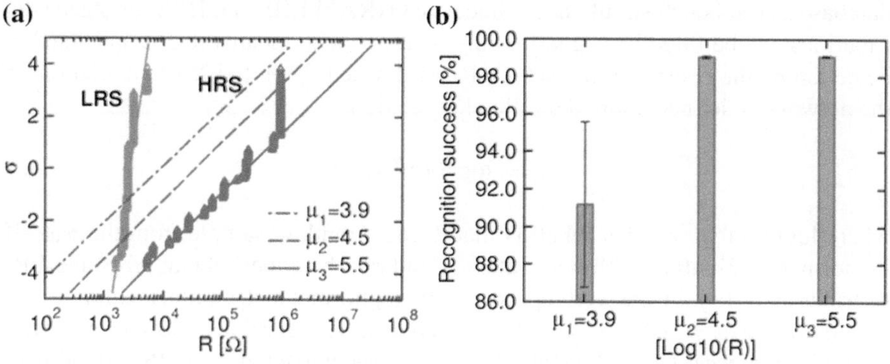

Fig. 14 **a** Distributions of LRS (*red*) and HRS (*blue*) for experimental 16 kbit (*symbols*) and HRS lognormal distributions implemented in our Xnet (*lines*) with different mean values (μ) and same variability (σ). **b** Impact of the mean value of the HRS lognormal distribution (μ) on recognition success of OxRAM-based CNN

4 Conclusions

In this chapter, we proposed a synapse that employs multiple OxRAM binary cells operating in parallel to achieve an analog behavior. The implementation of one VRRAM pillar per synapse was presented as a possible solution to gain area with respect to planar approaches. Moreover, we described a possible hardware implementation of a spike-based Convolutional Neural Network for visual pattern recognition using multiple binary OxRAM devices as synapses. The proposed solution, with respect to typical software implementation on CPUs and GPUs, allows to improve performances and to decrease the power consumption. Thanks to the use of the OxRAM synapses to implement the kernel, the convolution operations are performed directly in memory, in a fully parallel and distributed approach, allowing to achieve a latency of 1 μs per image (assuming a spike encoding frequency $f_{MAX} = 8$ MHz). The estimated latency per image for a software implementation of the network is 2.5 ms, using 16 parallel processing cores with a clock frequency of 200MHz. The impact of the synaptic resolution and OxRAM variability on the CNN performances were evaluated using the electrical data extracted from a 28 nm CMOS OxRAM array. The proposed CNN architecture is highly tolerant to variability. Recognition success rates higher than 99 and 97% have been demonstrated for the MNIST and GTSRB networks, respectively. These results are similar to the state-of-the-art recognition success rates obtained with formal CNN models, implemented with floating-point precision synapses. These success rates are reached using 11 and 20 OxRAM devices per synapse for the MNIST and GTSRB applications, respectively.

References

1. Mead, C.: Neuromorphic electronic systems. Proc. IEEE **78**, 1629–1636 (1990)
2. Indiveri, G.: Shih-Chii Liu: memory and information processing in neuromorphic systems. Proc. IEEE **103**, 1379–1397 (2015)
3. Prezioso, M., Merrikh-Bayat, F., Hoskins, B.D., Adam, G.C., Likharev, K.K., Strukov, D.B.: Training and operation of anintegrated neuromorphic network based on metal-oxide memristors. Nature **521**, 61–64 (2015)
4. Burr, G.W., Shelby, R.M., di Nolfo, C., Jang, J.W., Shenoy, R.S., Narayanan, P., Virwani, K., Giacometti, E.U., Kurdi, B., Hwang, H.: Experimental demonstration and tolerancing of a large-scale neural network (165,000 synapses), using phase-change memory as the synaptic weight element. In: Electron Devices Meeting (IEDM) IEEE International, pp. 29.5.1–29.5.4 (2014)
5. Kuzum, D., Yu, S., Wong, H.-S.P.: Synaptic electronics: materials, devices and applications. Nanotechnology **24**, 382001 (2013)
6. Suri, M., Bichler, O., Querlioz, D., Palma, G., Vianello, E., Vuillaume, D., Gamrat, C.: DeSalvo: CBRAM devices as binary synapses for lowpower stochastic neuromorphic systems: auditory (Cochlea) and visual (Retina) cognitive processing applications. In: Electron Devices Meeting (IEDM) IEEE International, pp. 10.3.1–10.3.4 (2012)
7. Kuzum, D., Jeyasingh, R.G.D., Wong,H.-S.P.: Energy efficient programming of nanoelectronic synaptic devices for large-scale implementation of associative and temporal sequence learning. In: Electron Devices Meeting (IEDM) IEEE International, pp. 30.3.1–30.3.4 (2011)
8. Suri, M., Bichler, O., Querlioz, D., Cueto, O., Perniola, L., Sousa, V., Vuillaume, D., Gamrat, C., DeSalvo, B.: Phase change memory as synapse for ultra-dense neuromorphic systems: application to complex visual pattern extraction. In: Electron Devices Meeting (IEDM) IEEE International, pp. 4.4.1–4.4.4 (2011)
9. Wang, I.-T., Lin, Y.-C., Wang, Y.-F., Hsu, C.-W., Hou, T.-H.: 3D synaptic architecture with ultralow sub-10 fJ energy per spike for neuromorphic computation. In: Electron Devices Meeting (IEDM) IEEE International, pp. 28.5.1–28.5.4 (2014)
10. Park, S., Sheri, A., Kim, J., Noh, J., Jang, J., Jeon, M., Lee, B., Lee, B.R., Lee, B.H., Hwang, H.: Neuromorphic speech systems using advanced ReRAM-based synapse. In: Electron Devices Meeting (IEDM) IEEE International, pp. 25.6.1–25.6.4 (2013)
11. Vianello, E., Thomas, O., Molas, G., Turkyilmaz, O., Jovanovic, N., Garbin, D., Palma, G., Alayan, M., Nguyen, C., Coignus, J., Giraud, B., Benoist, T., Reyboz, M., Toffoli, A., Charpin, C., Clermidy, F., Perniola, L.: Resistive memories for ultra-low-power embedded computing design. In: Electron Devices Meeting (IEDM) IEEE International, pp. 6.3.1–6.3.4 (2014)

12. Benoist, A., Blonkowski, S., Jeannot, S., Denorme, S., Damiens, J., Berger, J., Candelier, P., Vianello, E., Grampeix, H., Nodin, J.E., Jalaguier, E., Perniola, L., Allard, B.: 28nm advanced CMOS resistive RAM solution as embedded non-volatile memory. In: Proceedings of the IEEE Reliability Physics Symposium, pp. 2E.6.1–2E.6.5 (2014)
13. Govoreanu, B., Kar, G.S., Chen, Y., Paraschiv, V., Kubicek, S., Fantini, A., Radu, I.P., Goux, L., Clima, S., Degraeve, R., Jossart, N., Richard, O., Vandeweyer, T., Seo, K., Hendrickx, P., Pourtois, G., Bender, H., Altimime, L., Wouters, D.J., Kittl, J.A., Jurczak, M.: 10x10nm^2 Hf/HfOx crossbar resistive RAM with excellent performance, reliability and low-energy operation. In: Electron Devices Meeting (IEDM) IEEE International, pp. 31.6.1–31.6.4 (2011)
14. Garbin, D., Suri, M., Bichler, O., Querlioz, D., Gamrat, C., DeSalvo, B.: Probabilistic neuromorphic system using binary phase-change memory (PCM) synapses: detailed power consumption analysis. In: 3th IEEE Conference on Nanotechnology (IEEE-NANO), pp. 91–94 (2013)
15. Garbin, D., Vianello, E., Bichler, O., Rafhay, Q., Gamrat, C., Ghibaudo, G., DeSalvo, B., Perniola, L.: HfO_2-based OxRAM devices as synapses for convolutional neural networks. IEEE Trans. Electron Devices **62**, 2494–2501 (2015)
16. Bill, J., Legenstein, R.: A compound memristive synapse model for statistical learning through STDP in spiking neural networks. Front. Neurosci. **8**, 412 (2014)
17. Simard, P.Y., Steinkraus, D., Platt, J.C.: Best practices for convolutional neural networks applied to visual document analysis. In: Seventh IEEE International Conference on Document Analysis and Recognition, pp. 958–963 (2003)
18. Goldberg, D.H., Cauwenberghs, G., Andreou, A.G.: Probabilistic synaptic weighting in a reconfigurable network of VLSI integrate-and-fire neurons. Neural Netw. **14**, 781–793 (2001)
19. Piccolboni, G., Molas, G., Portal, J.M., Coquand, R., Bocquet, M., Garbin, D., Vianello, E., Carabasse, C., Delaye, V., Pellissier, C., Magis, T., Cagli, C., Gely, M., Cueto, O., Deleruyelle, O., Ghibaudo, G., DeSalvo, B., Perniola, L.: Investigation of the potentialities of Vertical Resistive RAM (VRRAM) for neuromorphic applications. In: lectron Devices Meeting (IEDM) IEEE International, pp. 17.2.1–17.2.4 (2015)
20. Garbin, D., Bichler, O., Vianello, E., Rafhay, Q., Gamrat, C., Perniola, L., Ghibaudo, G., DeSalvo, B.: Variability-tolerant convolutional neural network for pattern recognition applications based on OxRAM synapses. In: Electron Devices Meeting (IEDM) IEEE International, pp. 28.4.1–28.4.4 (2014)
21. Kim, S., Ishii, M., Lewis, S., Perri, T., BrightSky, M., Kim, W., Jordan, R., Burr, G.W., Sosa, N., Ray, A., Han, J.-P., Miller, C., Hosokawa, K., Lam, C.: NVM neuromorphic core with 64k-cell (256-by-256) phase change memory synaptic array with On-Chip neuron circuits for continuous in-situ learning. In: Electron Devices Meeting (IEDM) IEEE International, pp. 17.1.1–17.1.4 (2015)
22. Wang, Z., Ambrogio, S., Baletti, S., Ielemini, D.: A 2-transistor/1-resistor artificial synapse capable of communication and stochastic learning in neuromorphic systems. Front. Neurosci. **8**, 438 (2015)
23. Hubel, D.H., Wiesel, T.N.: Receptive fields, binocular interaction and functional architecture in the cats visual cortex. J. Physiol. **160**, 106–154 (1962)
24. Felleman, D.J., Van Essen, D.C.: Distributed hierarchical processing in the primate cerebral cortex. Cereb. Cortex **1**, 1–47 (1991)
25. Fukushima, K.: Artificial vision by multi-layered neural networks: neocognitron and its advances. Neural Netw. **37**, 103–119 (2013)
26. Ciresan, D., Meier, U., Masci, J., Schmidhuber, J.: Multi-column deep neural network for traffic sign classification. Neural Netw. **32**, 333–338 (2012)
27. Ning, F., Delhomme, D., LeCun, Y., Piano, F., Bottou, L., Barbano, P.E.: Toward automatic phenotyping of developing embryos from videos. IEEE Trans. Image Process. **14**, 1360–1371 (2005)
28. Sermanet, P., Kavukcuoglu, K., Chintala, S., LeCun, Y.: Pedestrian detection with unsupervised multi-stage feature learning. In: IEEE Conference on Computer Vision and Pattern Recognition (CVPR), p. 36263633 (2013)

29. Vaillant, R., Monrocq, C., LeCun, Y.: A convolutional neural network hand tracker. IEEE Proc.-Vision, Image. Signal Process. **141**, 245–250 (1994)
30. Nowlan, S.J., Platt, J.C.: Original approach for the localisation of objects in images. Adv. Neural Inf. Process. Syst. 901–908 (1995)
31. Garcia, C., Delakis, M.: Convolutional face finder: a neural architecture for fast and robust face detection. IEEE Trans. Pattern Anal. Mach. Intell. **26**, 14081423 (2004)
32. Taigman, Y., Yang, M., Ranzato, M., Wolf, L.: Deepface: closing the gap to human-level performance in face verification. In: IEEE Conference on Computer Vision and Pattern Recognition (CVPR), pp. 1701–1708 (2014)
33. Garbin, D., Vianello, E., Bichler, O., Azzaz, M., Rafhay, Q., Candelier, P., Gamrat, C., Ghibaudo, G., DeSalvo, B., Perniola, L.: On the impact of OxRAM-based synapses variability on convolutional neural networks performance. In: 2015 IEEE/ACM International Symposium on Nanoscale Architectures (NANOARCH), pp. 193–198 (2015)
34. LeCun, Y., Cortes, C., Burges, C.J.C.: The MNIST Database of Handwritten Digits. http://yann.lecun.com/exdb/mnist/
35. Stallkamp, J., Schlipsing, M., Salmen, J., Igel, C.: The German traffic sign recognition benchmark: a multi-class classification competition. In: Proceedings of the IEEE International Joint Conference on Neural Networks, pp. 1453–1460 (2011)
36. LeCun, Y., Bottou, L., Bengio, Y., Haffner, P.: Gradient-based learning applied to document recognition. Proc. IEEE **86**, 2278–2324 (1998)
37. Ciresan, D., Meier, U., Schmidhuber, J.: Multi-column deep neural networks for image classification. In: IEEE Conference on Computer Vision and Pattern Recognition (CVPR), pp. 3642–3649 (2012)
38. Bichler, O., Querlioz, D., Thorpe, S. J., Bourgoin, J.-P., Gamrat, C.: Unsupervised features extraction from asynchronous silicon retina through spike-timing-dependent plasticity. In: Proceedings of the IEEE International Joint Conference on Neural Networks, pp. 859–866 (2011)
39. Hayakawa, Y., Himeno, A., Yasuhara, R., Boullart, W., Vecchio, E., Vandeweyer, T., Witters, T., Crotti, D., Jurczak, M., Fujii, S., Ito, S., Kawashima, Y., Ikeda, Y., Kawahara, A., Kawai, K., Wei, K., Muraoka, S., Shimakawa, K., Mikawa, T., Yoneda, S.: Highly reliable TaO_x ReRAM with centralized filament for 28 nm embedded application. In: 2015 Symposium on VLSI Technology Digest of Technical Papers (2015)
40. Ueki, M., Takeuchi, K., Yamamoto, T., Tanabe, A., Ikarashi, N., Saitoh, M., Nagumo, T., Sunamura, H., Narihiro, M., Uejima, K., Masuzaki, K., Furutake, N., Saito, S., Yabe, Y., Mitsuiki, A., Takeda, K., Hase, T., Hayashi, Y.: Low-Power embedded ReRAM technology for IoT applications. In: 2015 Symposium on VLSI Technology Digest of Technical Papers (2015)
41. Vianello, E., Garbin, D., Jovanovic, N., Bichler, O., Thomas, O., Salvo, B., Perniola, L.: Oxide based resistive memories for low power embedded applications and neuromorphic systems. In: 2015 ECS Transactions (2015)

Nonvolatile Memory Crossbar Arrays for Non-von Neumann Computing

Severin Sidler, Jun-Woo Jang, Geoffrey W. Burr, Robert M. Shelby, Irem Boybat, Carmelo di Nolfo, Pritish Narayanan, Kumar Virwani and Hyunsang Hwang

Abstract In the conventional von Neumann (VN) architecture, data—both operands and operations to be performed on those operands—makes its way from memory to a dedicated central processor. With the end of Dennard scaling and the resulting slowdown in Moore's law, the IT industry is turning its attention to non-Von Neumann (non-VN) architectures, and in particular, to computing architectures motivated by the human brain. One family of such non-VN computing architectures is artificial neural networks (ANNs). To be competitive with conventional architectures, such ANNs will need to be massively parallel, with many neurons interconnected using a vast number of synapses, working together efficiently to compute problems of significant interest. Emerging nonvolatile memories, such as phase-change memory (PCM)

S. Sidler · I. Boybat
EPFL, Lausanne, Switzerland
e-mail: severin.sidler@epfl.ch

I. Boybat
e-mail: irem.boybat@epfl.ch

J.-W. Jang · H. Hwang
Pohang University of Science and Technology, Pohang, Korea
e-mail: junwoo410@postech.ac.kr

H. Hwang
e-mail: hwanghs@postech.ac.kr

G.W. Burr (✉) · R.M. Shelby · C. di Nolfo · P. Narayanan · K. Virwani
IBM Research, Almaden, San Jose, CA, USA
e-mail: gwburr@us.ibm.com

R.M. Shelby
e-mail: rshelby@us.ibm.com

C. di Nolfo
e-mail: cdinolfo@us.ibm.com

P. Narayanan
e-mail: pnaraya@us.ibm.com

K. Virwani
e-mail: kvirwan@us.ibm.com

© Springer (India) Pvt. Ltd. 2017

M. Suri (ed.), *Advances in Neuromorphic Hardware Exploiting Emerging Nanoscale Devices*, Cognitive Systems Monographs 31,
DOI 10.1007/978-81-322-3703-7_7

or resistive memory (RRAM), could prove very helpful for this, by providing inherently analog synaptic behavior in densely packed crossbar arrays suitable for on-chip learning. We discuss our recent research investigating the characteristics needed from such nonvolatile memory elements for implementation of high-performance ANNs. We describe experiments on a 3-layer perceptron network with 164,885 synapses, each implemented using 2 NVM devices. A variant of the backpropagation weight update rule suitable for NVM+selector crossbar arrays is shown and implemented in a mixed hardware–software experiment using an available, non-crossbar PCM array. Extensive tolerancing results are enabled by precise matching of our NN simulator to the conditions of the hardware experiment. This tolerancing shows clearly that NVM-based neural networks are highly resilient to random effects (NVM variability, yield, and stochasticity), but highly sensitive to gradient effects that act to steer all synaptic weights. Simulations of ANNs with both PCM and non-filamentary bipolar RRAM based on $Pr_{1-x}Ca_xMnO_3$ (PCMO) are also discussed. PCM exhibits smooth, slightly nonlinear partial-SET (conductance increase) behavior, but the asymmetry of its abrupt RESET introduces difficulties; in contrast, PCMO offers continuous conductance change in both directions, but exhibits significant nonlinearities (degree of conductance change depends strongly on absolute conductance). The quantitative impacts of these issues on ANN performance (classification accuracy) are discussed.

1 Introduction

Dense arrays of nonvolatile memory (NVM) and selector device pairs (Fig. 1) can implement neuro-inspired non-von Neumann computing [8, 16], using pairs [16] of NVM devices as programmable (plastic) bidirectional synapses. Prior work has emphasized the spike-timing-dependent plasticity (STDP) algorithm [8, 16], motivated by synaptic measurements in real brains. However, full experimental NVM demonstrations of successful network learning have been limited in size ($\leq$100 synapses), and few results have reported quantitative performance metrics such as classification accuracy. Worse yet, it has been difficult to be sure whether the relatively poor metrics reported to date might be due to immaturities or inefficiencies in the STDP learning algorithm (as it is currently implemented), rather than reflective of problems introduced by the imperfections of the NVM devices.

Unlike STDP, backpropagation is a widely used, well-studied method in training artificial neural networks, offering benchmark-able performance on datasets such as handwritten digits (MNIST) [11]. Although proposed earlier, it gained great popularity in the 1980s [11, 15], and with the advent of GPUs, backpropagation now dominates the neural network field. In the present work, we use backpropagation to train a relatively simple multilayer perceptron network (Fig. 2). During forward evaluation of this network, each layer's inputs (x_i) drive the next layer's neurons through weights w_{ij} and a nonlinearity $f()$ (Fig. 2). Supervised learning occurs (Fig. 3) by backpropagating error terms δ_j to adjust each weight w_{ij} as the second step. A 3-layer network is capable of accuracies, on previously unseen "test" images (*generaliza-*

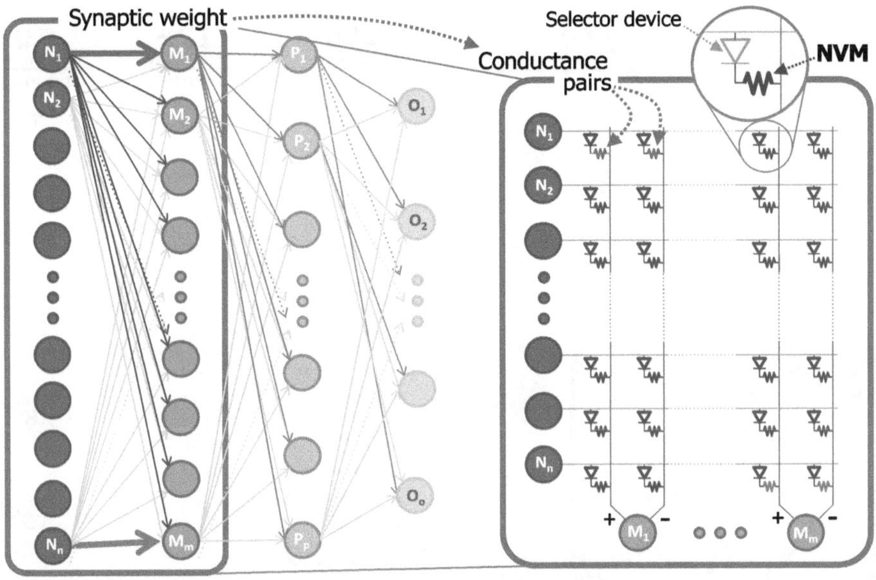

Fig. 1 Neuro-inspired non-Von Neumann computing [8, 16], in which neurons activate each other through dense networks of programmable synaptic weights, can be implemented using dense crossbar arrays of nonvolatile memory (NVM) and selector device pairs. (© IEEE, all rights reserved. Reprinted, with permission, from [4])

tion), of ~97 % [11] if all 60,000 training examples are used for training. (Fig. 4); even higher accuracy is possible by first "pre-training" the weights in each layer [7]. However, if fewer training images are used for training, higher training accuracy but lower generalization accuracy (94 %) on the "test" set is obtained (Fig. 4).

Here, we use tanh() as the nonlinear function f (), and one bias (always-ON) neuron is added to each layer other than the output layer, in addition to those neurons shown in Fig. 2. Like with STDP, low-power neurons should be achievable by emphasizing brief spikes [14] and local-only clocking. However, note that no CMOS neuron circuitry is built or even specified in this chapter—the focus here will be solely on the effects of the imperfections of the NVM elements.

We have chosen to work with phase-change memory (PCM) since we have access to large PCM arrays in hardware. We discuss the consequences of the fundamental asymmetry in PCM conductance response: The fact that small conductance increases can be implemented through "partial-SET" pulses, but the RESET (conductance decrease) operation tends to be quite abrupt. However, we also discuss the use of bidirectional NVM devices (such as non-filamentary RRAM [9]). We show that such a bidirectional NVM with a symmetric, linear conductance response is fully capable of delivering the same high classification accuracies (on the problem we study, handwritten digit recognition) as a conventional, software-based implementation of the same neural network.

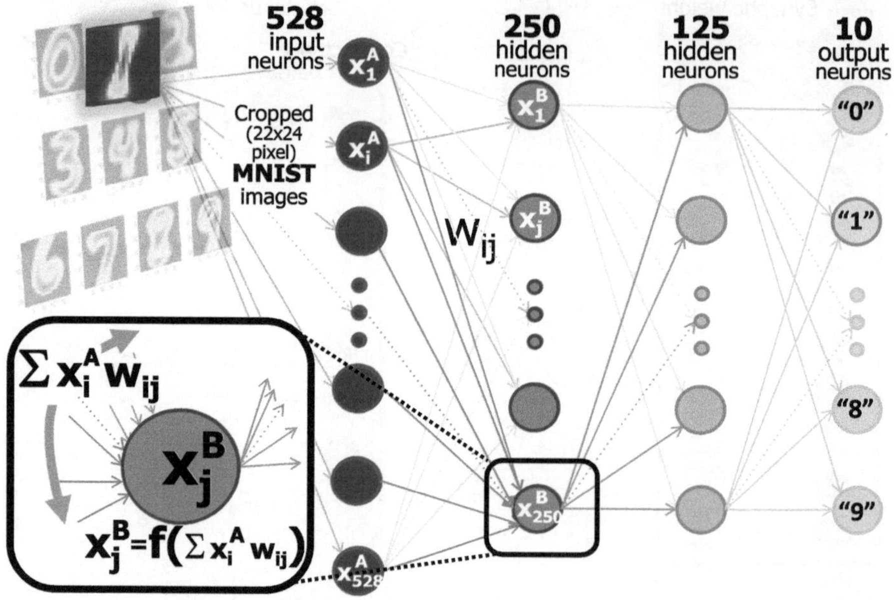

Fig. 2 In forward evaluation of a multilayer perceptron, each layer's neurons drive the next layer through weights w_{ij} and a nonlinearity $f()$. Input neurons are driven by pixels from successive MNIST images (cropped to 22×24); the 10 output neurons identify which digit was presented. (© IEEE, all rights reserved. Reprinted, with permission, from [4])

In Sect. 4, we describe additional simulation studies using the measured conductance response of a real bidirectional NVM: the non-filamentary RRAM based on $Pr_{1-x}Ca_xMnO_3$, also known as PCMO.

2 Considerations for a Crossbar Implementation

By encoding synaptic weight in the conductance difference between a pair of nonvolatile memory devices, $w_{ij} = G^+ - G^-$ [16], forward propagation simply compares total read signal on columns (Fig. 5). (A similar parallel read operation on rows enables the backpropagation of δ corrections.)

This can be performed by encoding x using some combination of voltage-domain or time-domain encoding (either number of read pulses, or pulse duration, or some appropriate combination of both). These CMOS circuitry choices are interesting and important topics, but are beyond the scope of this chapter. Any nonvolatile memory device that can offer a nondestructive parallel read (as shown in Fig. 5) of memory states that can be smoothly adjusted up or down through a wide range of analog values could potentially be used in this application. Here, we focus on NVM devices that offer a range of analog conductance states.

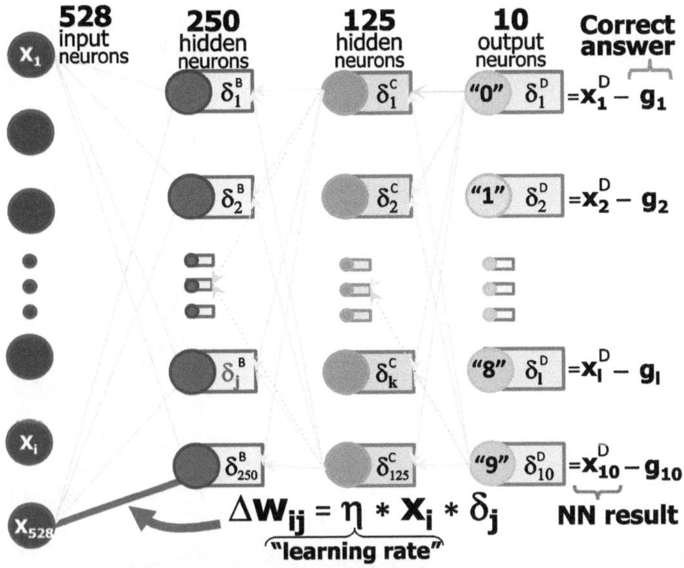

Fig. 3 In supervised learning, error terms δ_j are backpropagated, adjusting each weight w_{ij} to minimize an "energy" function by gradient descent, reducing classification error between computed (x_l^D) and desired output vectors (g_l). (© IEEE, all rights reserved. Reprinted, with permission, from [4])

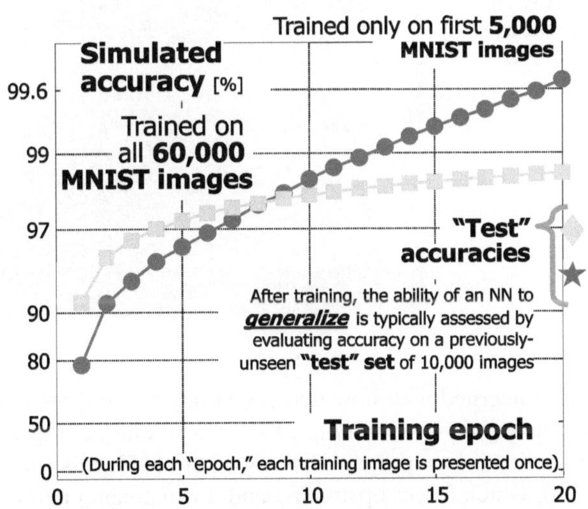

Fig. 4 A 3-layer perceptron network can classify previously unseen ("test") MNIST handwritten digits with up to ~97 % accuracy [11]. Training on a subset of the images sacrifices some generalization accuracy but speeds up training. (© IEEE, all rights reserved. Reprinted, with permission, from [4])

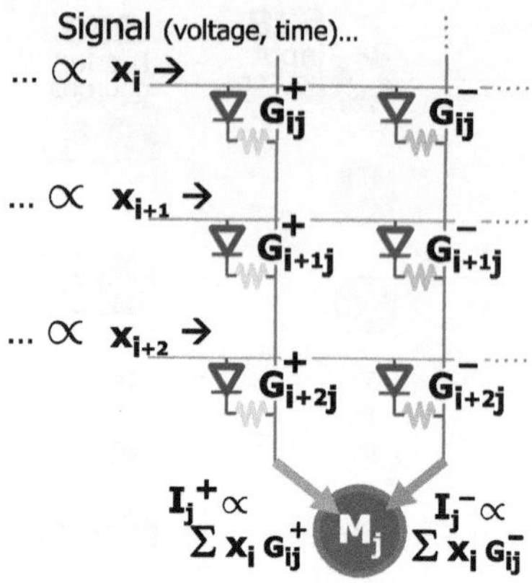

Fig. 5 By comparing total read signal between pairs of bitlines, summation of synaptic weights (encoded as conductance differences, $w_{ij} = G^+ - G^-$) is highly parallel. (© IEEE, all rights reserved. Reprinted, with permission, from [4])

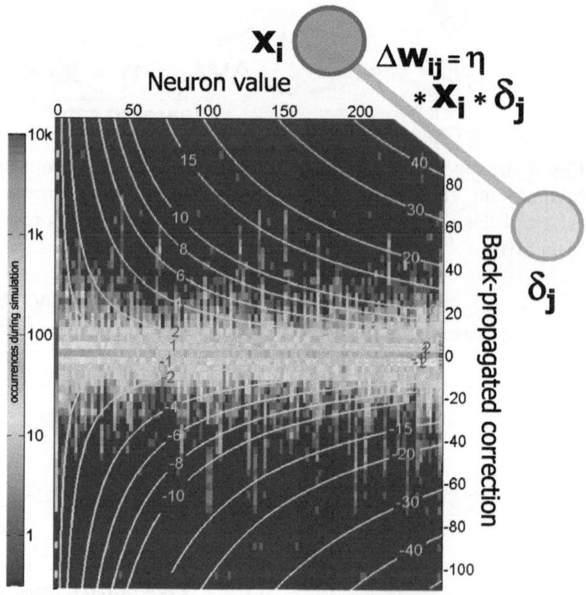

Fig. 6 Backpropagation calls for each weight to be updated by $\Delta w_{ij} = \eta\, x_i\, \delta_j$, where η is the *learning rate*. Colormap shows log(occurrences), in the 1st layer, during NN training (*blue curve*, Fig. 4); white contours identify the quantized increase in the integer weight. (© IEEE, all rights reserved. Reprinted, with permission, from [4])

This chapter is concerned with how real NVM devices will respond to programming instructions during in situ training of their artificial neural network. Unfortunately, the conventional backpropagation algorithm [15] calls for weight updates $\Delta w_{ij} \propto x_i \delta_j$ (Fig. 6), which forces upstream i and downstream j neurons to exchange information uniquely for each and every synapse. This serial, element-by-element information exchange between neurons is highly undesirable in a crossbar array implementation. One alternative is to have each neuron, downstream and upstream, fire pulses based on their local knowledge of x_i and δ_j, respectively. The presence of a nonlinear selector is critical to ensure that NVM programming occurs only when pulses from both the upstream and downstream neurons overlap. This allows neurons

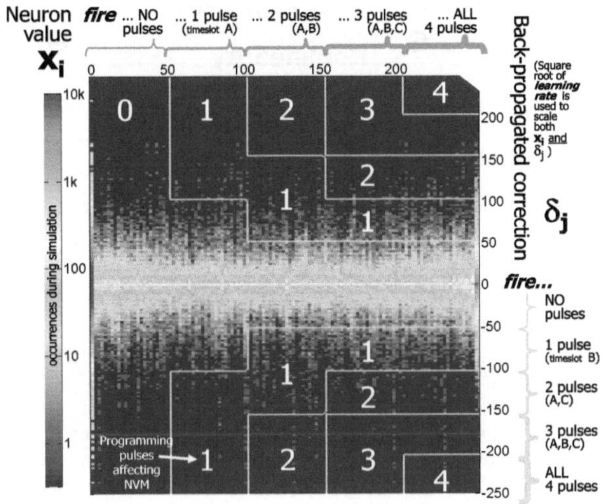

Fig. 7 In a crossbar array, efficient learning requires neurons to update weights in parallel, firing pulses whose overlap at the various NVM devices implements training. Colormap shows log(occurrences), in the 1st layer, during NN training (*red curve*, Fig. 8); *white contours* identify the quantized increase in the integer weight. (© IEEE, all rights reserved. Reprinted, with permission, from [4])

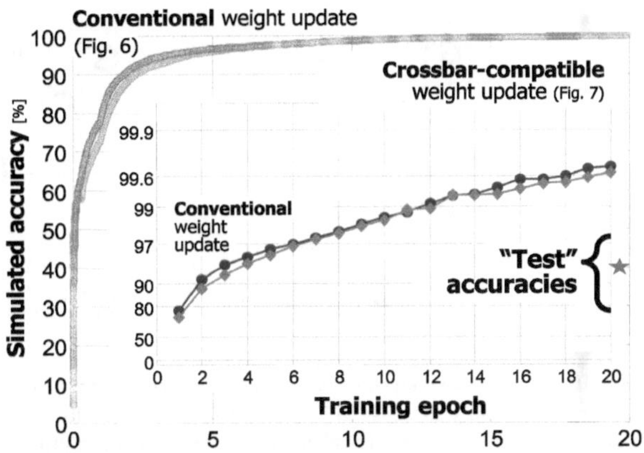

Fig. 8 Computer NN simulations show that a crossbar-compatible weight update rule (Fig. 7) is just as effective as the conventional update rule (Fig. 6). (© IEEE, all rights reserved. Reprinted, with permission, from [4])

to modify weights in parallel, making learning much more efficient [8] (Fig. 7). (Note that to reduce peak power, one might choose to stagger these write pulses across the array, one sub-block at a time.) Figure 8 shows, using a simulation of the neural network in Figs. 2 and 3, that this adaptation for nonvolatile memory implementation has no adverse effect on accuracy.

However, while modifying the update rule is clearly not a problem, the conductance response of any real nonvolatile memory device exhibits imperfections that can decidedly affect the neural network performance. These imperfections include nonlinearity, stochasticity, varying maxima, asymmetry between increasing/decreasing responses, and non-responsive devices at low or high conductance (Fig. 9). The initial version of this work [4] was the first paper to study the relative importance of each of these factors. A later, expanded version added significantly more explanatory details, adding several new plots detailing paths for future improvement [5]. A recently published conference paper included both a summary of this work as well

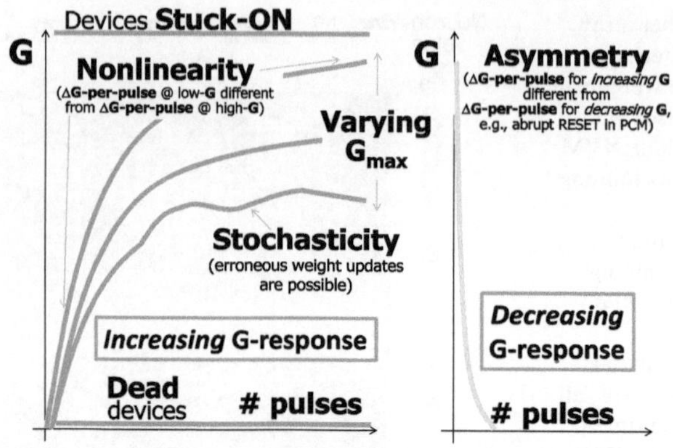

Fig. 9 The conductance response of an NVM device exhibits imperfections, including nonlinearity, stochasticity, varying maxima, asymmetry between increasing/decreasing responses, and non-responsive devices (at low or high G). (© IEEE, all rights reserved. Reprinted, with permission, from [4])

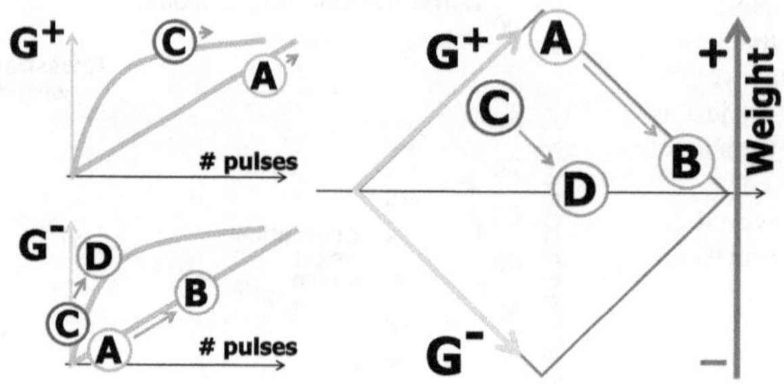

Fig. 10 If G values can only be increased (asymmetric G-response), a synapse at point (**A**) (G^+ saturated) can only increase G^-, leading to a low weight value (**B**). If response at small G values differs from that at large G (nonlinear G-response), alternating weight updates can no longer cancel. As synapses tend to get herded into the same portion of the G-diamond (**C** $\rightarrow$ **D**), the decrease in average weight can lead to network freeze-out. (© IEEE, all rights reserved. Reprinted, with permission, from [4])

as an initial assessment of the potential improvements in power and speed offered by NVM-based acceleration of machine learning [3].

The nonlinearity and asymmetry in the G-response can strongly degrade accuracy [4]. The "G-diamond"—a diamond-shaped plot of G^+ versus G^- in which weight is vertical position—is a graphical method for illustrating the synaptic "state" of a nonvolatile memory pair. In PCM-based synapses, the G-response is highly *asymmetric* and only partial-SET can be done gradually. In this context (Fig. 10), the synapse state can only move unidirectionally, from left to right, on the G-diamond. Bipolar filamentary RRAM such as HfO_x or TaO_x [18] or CBRAM [17] has a similar problem, except that the names are reversed: SET is the abrupt step and it is the RESET step which can be performed gradually.

Once one G value is saturated, subsequent training can only increase the other G value, reducing weight magnitude. *Nonlinearity* in G-response further encourages weights of low value. If the response at small G values differs from that at large G, alternating weight updates no longer cancel. As synapses are herded into the same portion of the G-diamond (Fig. 10), the decrease in average weight can lead to network "freeze-out." In such a condition, the network chooses to update very few if any weights, meaning that the network stops evolving toward higher accuracy. Worse yet, since the few weight updates that do occur are quite likely to lead to weight magnitude decay, previously trained information is steadily erased and accuracy can actually decrease [4].

3 Phase-Change Memory (PCM): Results

One solution to the highly asymmetric response of PCM devices is occasional RESET [16], moving synapses back to the left edge of the "G-diamond" while preserving weight value (using an iterative SET procedure, Fig. 11 inset). However, if this is not done frequently enough, weight stagnation will degrade neural network accuracy (Fig. 11). (An analogous approach for bipolar filamentary RRAM or CBRAM would be occasional SET.)

3.1 Experimental Results

We implemented a 3-layer perceptron of 164,885 synapses (Figs. 2 and 3) on a 500 × 661 array of mushroom cell [2], 1T1R PCM devices (180 nm node, Fig. 12). While the

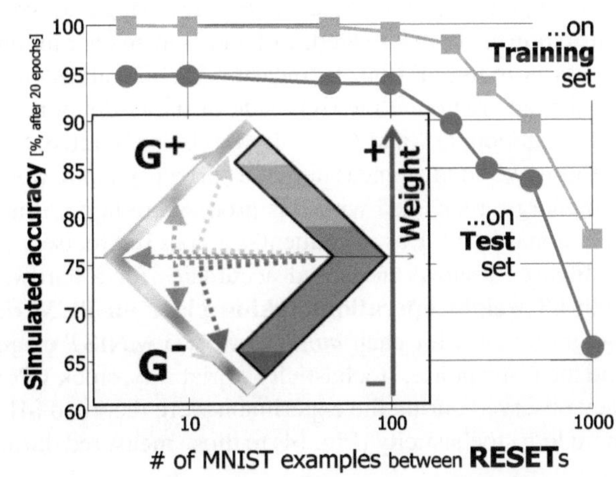

Fig. 11 Synapses with large conductance values (inset, *right edge* of G-diamond) can be refreshed (*moved left*) while preserving the weight (to some accuracy) by RESETs to both G followed by a partial-SET of one. If such RESETs are too infrequent, weight evolution stagnates and NN accuracy degrades. (© IEEE, all rights reserved. Reprinted, with permission, from [4])

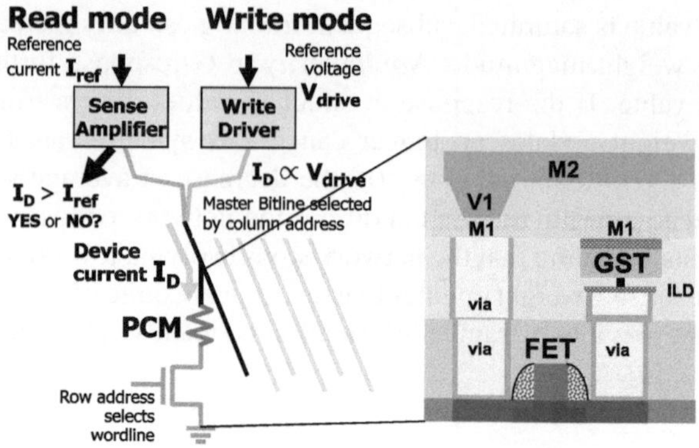

Fig. 12 Mushroom cell [2], 1T1R PCM devices (180 nm node) with 2 metal interconnect layers enable 512 × 1024 arrays. A 1-bit sense amplifier measures G values, passing the data to software-based neurons. Conductances are increased by identical 25 ns "partial-SET" pulses to increase G^+ (G^-) (Fig. 7), or by RESETs to both G followed by an iterative SET procedure (Fig. 11). (© IEEE, all rights reserved. Reprinted, with permission, from [4])

weight update algorithm (Fig. 7) is fully compatible with a crossbar implementation, our hardware allows only sequential access to each PCM device (Fig. 12). For read, a sense amplifier measures G values, passing the data to software-based neurons. Although this measurement is performed sequentially, weight summation and weight update procedures in the software-based neurons closely mimic the column- and row-based integrations. (Again, since no particular CMOS circuitry has been specified, we assume that the 8-bit value of x_i is implemented completely accurately. Any problems introduced by inaccurate encoding of x_i values by real CMOS hardware could be easily assessed using our tolerancing simulator.)

Weights are increased (decreased) by identical "partial-SET" pulses (Fig. 7) to increase G^+ (increase G^-) [4]. The deviation from true crossbar implementation occurs upon occasional RESET (Fig. 11), triggered when either G^+ or G^- are large, thus requiring both knowledge of and control over individual G values. Serial device access is required, both to measure the G values (to determine which are in the "L-shaped" region at the right side of the G-diamond) and then to fire two RESET pulses (at both G^+ and G^-) followed by an iterative SET procedure to increase one of those two conductances until the correct synaptic weight is restored. Since the time and energy associated with this process are large, it is highly desirable to perform occasional RESET as infrequently and as imprecisely as possible.

Figure 13 shows measured accuracies for a hardware synapse neural network, with **all weight operations taking place on PCM devices**. To reduce test time, weight updates for each *mini-batch* of 5 MNIST examples were applied together, and the G-response, stochasticity, variability, stuck-ON pixel rate, and RESET accuracy observed during the experiment were recorded [4]. By matching all parameters including stochasticity (Fig. 14) to those measured during the experiment, our neural

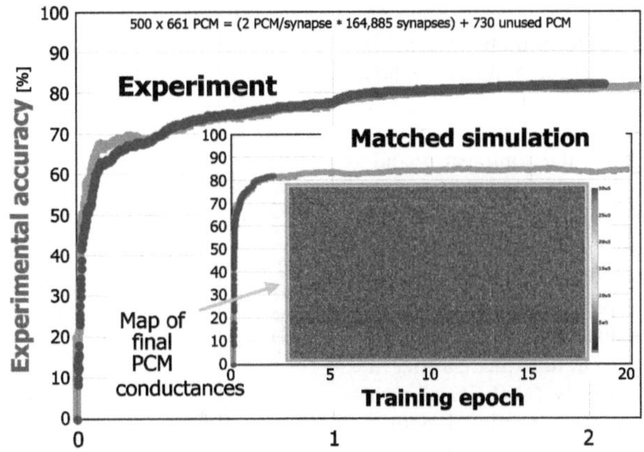

Fig. 13 Training and test accuracy for a 3-layer perceptron of 164,885 hardware synapses, with all weight operations taking place on a 500 × 661 array of mushroom cell [2] PCM devices (Fig. 12). Also shown is a matched computer simulation of this NN, using parameters extracted from the experiment. (© IEEE, all rights reserved. Reprinted, with permission, from [4])

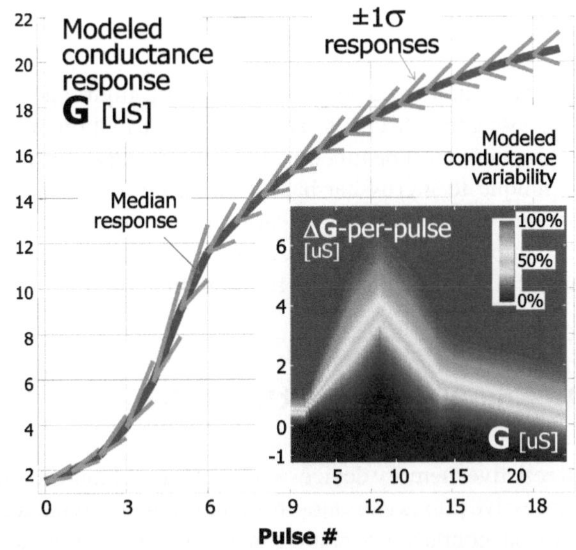

Fig. 14 Fitted G-response versus number of pulses (*blue* average, *red* $\pm 1\sigma$ responses) obtained from our computer model (inset) for the rate and stochasticity of G-response (ΔG per pulse vs. G) matched to experiment [4]. (© IEEE, all rights reserved. Reprinted, with permission, from [4])

network computer simulation was able to precisely reproduce the measured accuracy trends (Fig. 13).

We could then use that matched neural network simulation to explore the importance of nonvolatile memory imperfections. Final training (test) accuracy was toleranced as a function of variations in nonvolatile memory and neural network parameters away from the conditions used in our hardware demo [4]. NN performance was found to be highly robust to stochasticity, variable maxima, the presence of non-responsive devices, and infrequent and inaccurate RESETs [4]. However, as mentioned earlier, nonlinearity and asymmetry in G-response limited the maximum possible accuracy to ~85 %, and required precise tuning of the learning rate and neuron response (f') (see blue and red curves in Fig. 15). Too low a learning rate

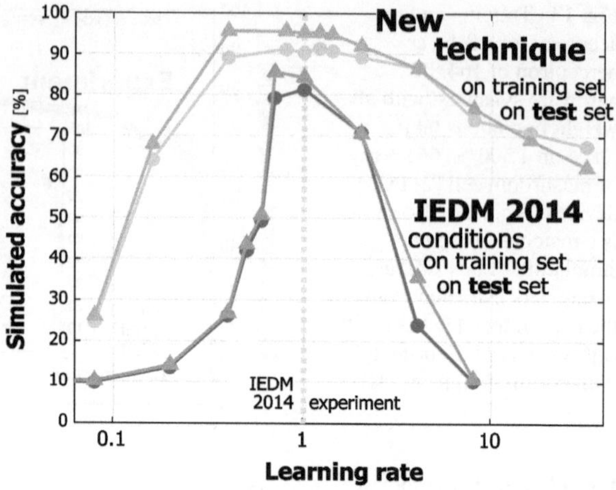

Fig. 15 A large number of synapses tend to "dither," with frequent updates whose aggregate effect *ought* to be zero (but which is nonzero due to the nonlinearity and asymmetry of NVM-based synapses). By suppressing update of such synapses, NN performance can be improved and training energy reduced, while reducing the need to tune the learning rate precisely. (© IEEE, all rights reserved. Reprinted, with permission, from [3].)

and no weight receives any update; too high, and the imperfections in the NVM response generate chaos. The narrow distribution of these parameters means that the experiment must be tuned very carefully. An extension of an existing neural network technique to a crossbar-based neural network has been found to provide a much broader distribution of the learning rate (magenta and cyan curves in Fig. 15), while also improving overall performance. This technique is currently under investigation and will be the subject of a future publication.

4 Non-filamentary RRAM Results

In resistive memory devices such as phase-change memory (PCM), the application of successive pulses can smoothly change analog conductance in one direction (increasing), but conductance change in the other direction (decreasing) is regrettably abrupt, returning to the conductance extrema after a single pulse. Resistive memory devices that offer bidirectional analog conductance change [9, 10] could potentially lead to more power- and area-efficient systems.

Resistive memory technology is a promising synaptic device due to its analog memory characteristics offering many intermediate conductance states, the high density of the cross-point array structure, and low power consumption.

In this section, we study the optimum potentiation and depression characteristics of bidirectional synaptic devices for neuromorphic systems. To investigate the effects of the potentiation and depression characteristics on the system, we simulate the same 3-layer multilayer perceptron described earlier for the application of handwritten digit classification, using the measured bidirectional switching characteristics of $Pr_{1x}Ca_xMnO_3$ (PCMO)-based synaptic devices.

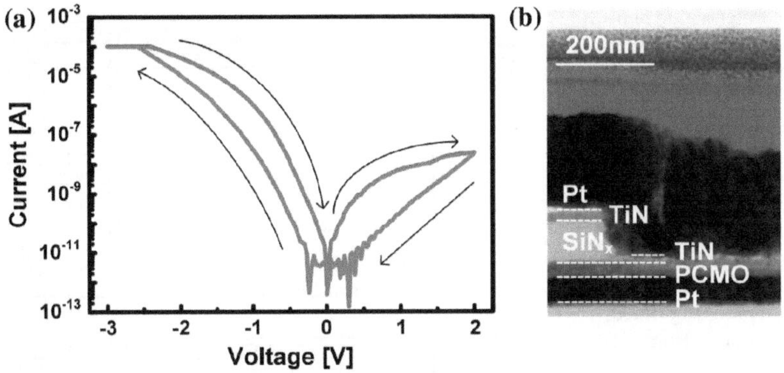

Fig. 16 **a** Current–voltage characteristics of TiN/PCMO-based resistive memory and **b** a TEM image of the device. (© IEEE, all rights reserved. Reprinted, with permission, from [9])

4.1 Fabrication of PCMO Devices

We fabricated 1 k-bit PCMO-based resistive memory arrays for evaluation as synaptic devices (Fig. 16), extending upon earlier work [13]. For device fabrication, a 50-nm-thick Pt layer for a bottom electrode and a 30-nm-thick polycrystalline PCMO film were deposited and patterned using conventional lithography and reactive ion etching. Next, an 80-nm-thick SiN_x layer was deposited by chemical vapor deposition, and via-holes (ranging in size from 0.15 to 1.0 μm) were formed by conventional lithography and reactive ion etching. A 10-nm-thick TiN layer and an 80-nm-thick Pt layer for a top electrode were deposited and patterned by conventional lithography. Electrical characteristics of the resistive memory devices based on PCMO were measured using an Agilent B1500A (Fig. 17). Reads were performed at 1.0 V; write currents ranged from ~0.1 nA to ~1.0 mA.

Based on the measured device characteristics, we performed simulations of a neural network with three layers of synapses (Fig. 1) using the backpropagation algorithm [15].

Previously, we proposed a resistive memory-based synaptic device model (1) for various potentiation and depression characteristics to find conductance change behavior which can optimize the performance of a neuromorphic system [9].

$$G = \begin{cases} ((G_{LRS}^{\alpha} - G_{HRS}^{\alpha}) \times w + G_{HRS}^{\alpha})^{1/\alpha} & \text{if } \alpha \neq 0 \\ G_{HRS} \times (G_{LRS}/G_{HRS})^{w} & \text{if } \alpha = 0 \end{cases}, \tag{1}$$

where G_{LRS} and G_{HRS} are low-resistance state (LRS) and high-resistance state (HRS) conductance, respectively, α is a parameter that controls potentiation (α_p) or depression (α_d) characteristics, and w is an internal variable which ranges from 0 to 1. During learning, w increases or decreases as potentiating (depressing) pulses are applied to the resistive memory- based synaptic device. The potentiation and depres-

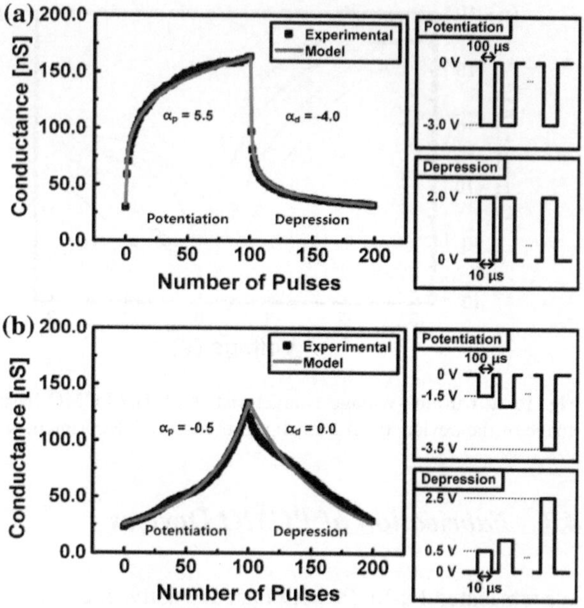

Fig. 17 Potentiation and depression characteristics of the experimental TiN/PCMO resistive memory and the proposed resistive memory-based synaptic device model when **a** identical pulses are applied and **b** nonidentical (increasing amplitude) pulses are applied. (© IEEE, all rights reserved. Reprinted, with permission, from [9])

sion characteristics of the resistive memory-based synaptic device model are concave down if $\alpha > 1$, and concave up if $\alpha < 1$.

4.2 Simulation Results

To investigate the effect of α on the neuromorphic system, we evaluated neural network accuracies as both α_p and α_d were varied. The fixed, unit-less G_{LRS} and G_{HRS} values (64 and 319), and the size of the smallest change in w (0.004), were based on the measurement data (Fig. 17), resulting in an on–off ratio of 5 and 256 effective multiple-conductance levels. From these simulations, we expect the conductance response in these PCMO devices to lead to a classification "test" accuracy of 82.38 % when identical pulses are used (Fig. 18b) and 90.55 % when nonidentical pulses are used (Fig. 18c). The highest possible accuracy, which occurs when the switching behavior is perfectly linear and symmetric (Fig. 18d), is 94.31 %. (At this point, the only remaining non-ideality is the constraint on weight magnitude imposed by the finite G_{LRS} [4].) However, similarly high accuracies, 88.93 % (Fig. 18e) and 86.12 % (Fig. 18f), can be obtained for nonlinear conductance responses, so long as the increasing and decreasing conductance responses are mirror-symmetric.

The G-diamond plots discussed earlier [4] can be used to represent distributions of G^+ and G^- in the neural network. Such plots represent both conductance values together with the resulting synaptic weight, $G = G^+ - G^-$ (as vertical position within the diamond) [4]. The weights, initially distributed uniformly along the center

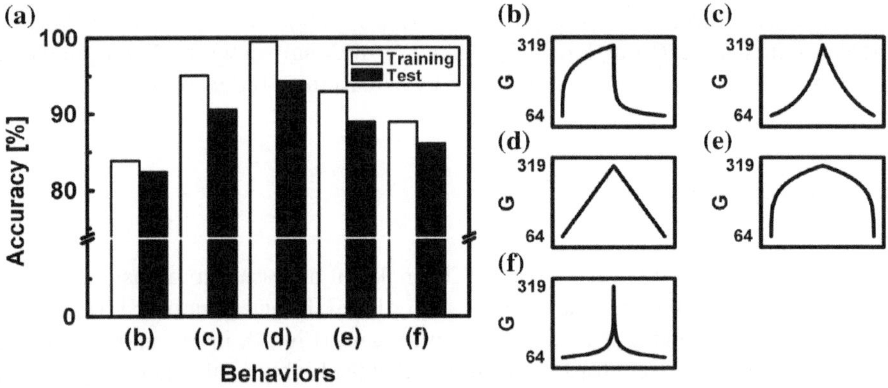

Fig. 18 **a** Calculated classification accuracies when **b** identical pulses are used ($\alpha_p = 5.5$, $\alpha_d = 4.0$), **c** nonidentical pulses are used ($\alpha_p = 0.5$, $\alpha_d = 0.0$), **d** the conductance-change behavior is perfectly linear and symmetric ($\alpha_p = \alpha_d = 1.0$), **e** and **f** the conductance-change behaviors are nonlinear but are mirror-symmetric ($\alpha_p = \alpha_d = 5.5$, and $\alpha_p = \alpha_d = 4.0$, respectively). (© IEEE, all rights reserved. Reprinted, with permission, from [9])

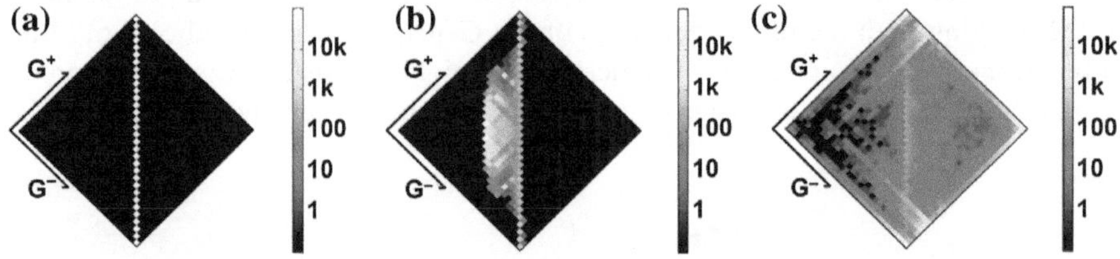

Fig. 19 Diamond-shaped plots of G^+ versus G (weight is vertical position [4]) **a** before training, and **b** after training when identical pulses are used ($\alpha_p = 5.5$, $\alpha_d = 4.0$) or **c** after training when nonidentical pulses are used ($\alpha_p = 0.5$, $\alpha_d = 0.0$). (© IEEE, all rights reserved. Reprinted, with permission, from [9])

axis of the G-diamond (Fig. 19a), spread out during neural network training. When identical pulses are used, the resulting G^+ and G^- values tend to concentrate around low weight values (Fig. 19b), preventing the neural network from utilizing the full range of possible weights. On the other hand, when nonidentical pulses are used, G^+ and G are spread out more (Fig. 19c), allowing the neural network to utilize the full range of weights.

However, the use of nonidentical training pulses in such a neuromorphic system requires additional external circuits, because the system has to read the conductance of a synaptic device before programming it, in order to identify which nonidentical training pulse to apply. Thus, there is a trade-off between the higher accuracy, and the resulting lower chip-area efficiency, higher power, and longer training time associated with the need to repeatedly measure individual conductances.

Total power is difficult to estimate without specifying the CMOS circuitry. With our current devices, training power would certainly be dominated by the large PCMO write energy (currently, 300 nJ/pulse = (100 μs)(3 V)(1 mA)). However, further increases in scaling (from 1 μm diameter or ~8e5 nm^2 down to 20 nm diameter ~300 nm^2) can be expected to reduce this by at least three orders of magnitude [12].

5 Discussion

While the asymmetric G-response of PCM makes it necessary to occasionally stop training, measure all conductances, and apply RESETs and iterative SETs, energy usage can be reasonable if RESETs are infrequent (Fig. 20, inset), and if learning rate is low (Fig. 20). In contrast, we observed that the highly nonlinear G-response of PCMO devices (Fig. 17a) degrades accuracy, unless additional time and energy are spent to identify the conductance states prior to selecting an appropriate programming pulse (Fig. 17b).

Neural networks based on bidirectional nonvolatile memory-based synapses can deliver high classification accuracy if G-response is linear and symmetric rather than nonlinear. We have previously explored the trends with an ideal but nonlinear NVM, varying both the initial steepness of the G-response and the choice of "fully bidirectional" weight updates (when increasing weight, for instance, we both increase G^+ **and** decrease G^- together) or "alternating bidirectional" (we choose one, but

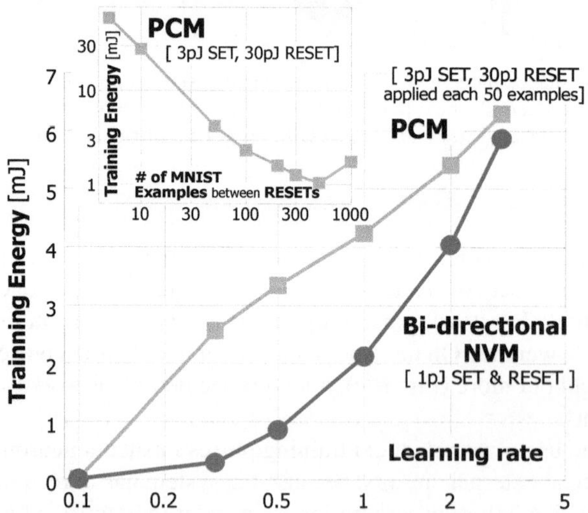

Fig. 20 Despite the higher power involved in RESET rather than partial-SET (30 pJ and 3 pJ for highly scaled PCM [8]), total energy costs of training can be minimized if RESETs are sufficiently infrequent (inset). Low-energy training requires low learning rates, which minimize the number of synaptic programming pulses. At higher learning rates, even a bidirectional, linear NVM requiring no RESET and offering low power (1 pJ per pulse) can lead to large training energy. (© IEEE, all rights reserved. Reprinted, with permission, from [4])

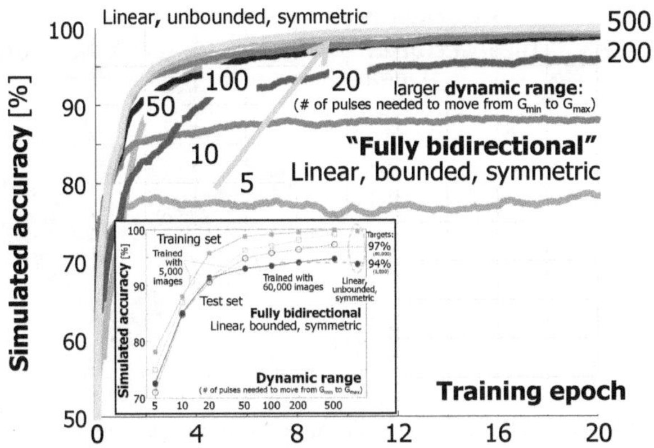

Fig. 21 NN performance (classification accuracy during training) when updating both G^+ and G^- ("fully bidirectional" scheme), with a linear G-response. The inset shows that when the dynamic range of the linear response is large, the classification accuracy can now reach that of the original network (a *test* accuracy of 94 % when trained with 5,000 images; 97 % when trained with all 60,000 images). (© IEEE, all rights reserved. Reprinted, with permission, from [5])

not both, of these two steps) [5]. A less steep response was found to be favorable, and the distinction between fully or alternating bidirectional has the most impact for steeply nonlinear G-responses [5].

The most ideal NVM, with a linear and symmetric conductance response in both directions, would result in more regularly distributed weight values and less freeze-outs, leading to higher accuracies. In Fig. 21, we show that a gentle linear response (e.g., a large number of identical pulses are needed to change the conductance from minimum to maximum conductance and vice versa) is advantageous compared to a steep response. While both the alternating bidirectional and fully bidirectional update schemes deliver higher accuracies than an NVM with a nonlinear conductance response, only the fully bidirectional update scheme reaches the same high test accuracies exhibited by networks in which the NVM conductances are unbounded (Fig. 21, inset).

The reason for this difference is that when the state of the synapse is at the boundaries of the G-diamond, there is a significant chance that the next weight update using the alternating bidirectional scheme will have little or no impact, simply because a conductance that is already saturated cannot be increased (decreased) any further. In the fully bidirectional update scheme, some amount of weight update will still occur at the edges of the G-diamond, leading to smaller discrepancies between the desired and actual weight changes, and thus higher performance. In addition, because the weights only move "up" and "down" the G-diamond in the fully bidirectional scheme, the synapses stay in the center stripe of the G-diamond (Fig. 22b), where they have access to the full dynamic range available. In contrast, because each weight update in the alternating bidirectional scheme moves along a

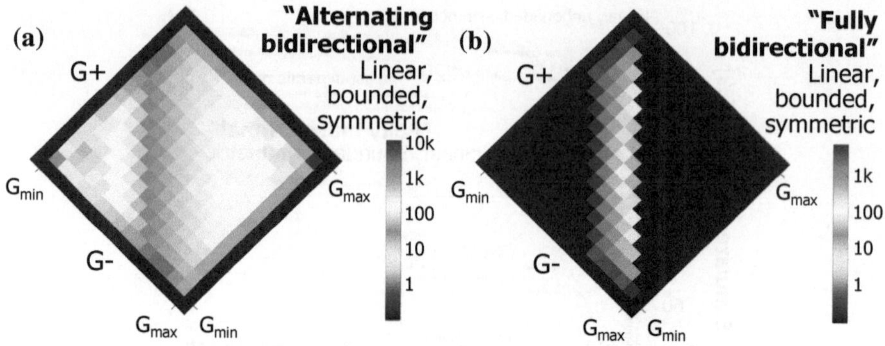

Fig. 22 When the G-response is steeply nonlinear, a "fully bidirectional" scheme exhibits lower accuracy [4] because any single weight update could potentially make two overly large conductance changes instead of just one. However, the "fully bidirectional" scheme provides better performance for a linear response with high dynamic range (see Fig. 21), because the small symmetric changes of each conductance move the synaptic weight up and down along the central *vertical axis* of the G-diamond. In contrast, the "alternating bidirectional" scheme can move some synapses to the *left* or *right* edges of the G-diamond, where the effective dynamic range (maximum weight magnitude) is significantly reduced. (© IEEE, all rights reserved. Reprinted, with permission, from [5])

diagonal line, some number of synapses end up at the edges of the G-diamond, where the effective dynamic range which they can access is significantly reduced (Fig. 22a).

These results demonstrate conclusively that NVM devices should be fully capable of delivering the same classification accuracy on the MNIST handwritten digits as a conventional implementation of this artificial neural network. All that is required of the NVM device is that it offers a bidirectional, linear, and symmetric response in conductance with large dynamic range (e.g., the change due to any one pulse represents only a small fraction of the entire conductance range available).

Figure 23 compares predicted training time (per ANN example) and power for two configurations of PCM-based on-chip machine learning against conventional GPU training. Under aggressive assumptions for parallel-read and parallel-write speed, PCM-based on-chip machine learning can potentially offer lower power and faster training for both large and small networks [3]. However, as circuit sharing, c_s, increases, the speed benefits of PCM-based on-chip machine learning disappear, since significant time is spent re-reading different columns of the same arrays because of the lack of sufficient readout parallelism [3]. The additional time and energy associated with the "occasional RESET" required for PCM devices are included here, and remain modest so long as this step is sufficiently infrequent [3].

Other future work will be needed to demonstrate a full crossbar-array implementation, including dedicated CMOS circuitry for summation of synaptic weights during both forward and backpropagation through nearly identical high-performance nonlinear selector devices. The values of neurons (x) and backpropagated errors (δ) will need to be stored in CMOS circuitry and presented to the crossbar, through some combination of analog voltage levels, number of read pulses, and/or dura-

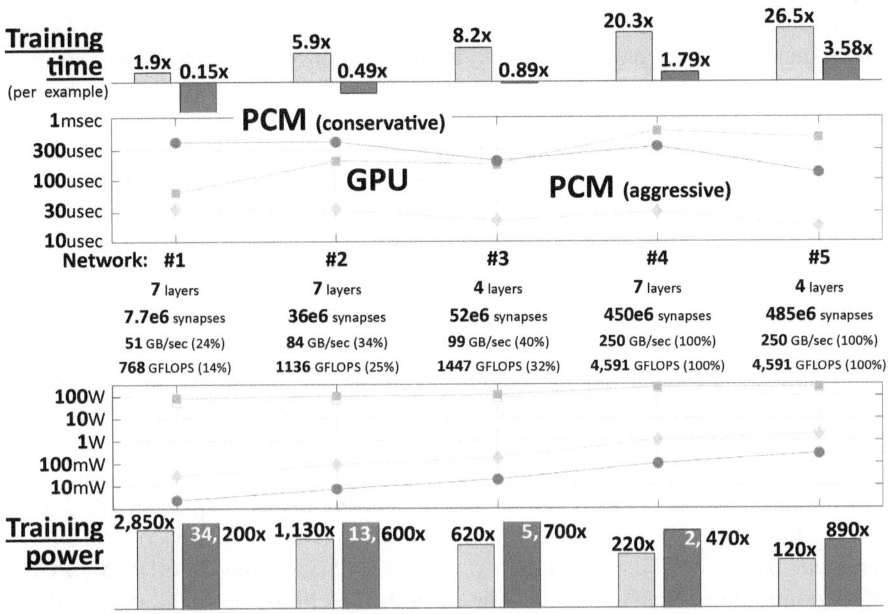

Fig. 23 Predicted training time (per ANN example) and power for 5 ANNs, ranging from 0.2 GB to nearly 6 GB [3]. Network #1 assumes Tesla K10 w/o momentum [1, 6]; #2–#5, Tesla K20x and momentum [1]. Mini-batch size is 1024, RESET interval $\mathbf{R} = 200$, circuit sharing $\mathbf{c}_s = 4$, *solid* (*dotted*) line assumes 50 W (20 W) idle GPU power. Under aggressive assumptions for parallel-read and parallel-write speed, PCM-based on-chip machine learning could potentially offer lower power and faster training for both large and small networks. (© IEEE, all rights reserved. Reprinted, with permission, from [3])

tion of read pulses. The need to synchronize write pulse timing between upstream and downstream neurons, and techniques to disperse the high-energy writes in time (to reduce the load on write drivers and voltage supplies), must also be addressed in future work, while maintaining sufficient speedup over existing GPU-based solutions. Neural network performance (test or generalization accuracy) must be increased to levels competitive with CPU- or GPU-based solutions, both on the MNIST dataset as well as newer and larger datasets.

6 Conclusions

Using two phase-change memory (PCM) devices per synapse, a 3-layer perceptron with 164,885 synapses was trained with backpropagation on a subset (5000 examples) of the MNIST database of handwritten digits to high accuracy of 82.2 % on the training set and 82.9 % on the test set. A weight update rule compatible for NVM+selector crossbar arrays was developed and was shown to have no adverse effect on accuracy. A novel "*G*-diamond" concept (Fig. 10) was introduced to illustrate issues created by nonlinearity and asymmetry in NVM conductance response. Asymmetry can be mitigated by an occasional RESET strategy, which can be both infrequent and inaccurate.

Using a neural network (NN) simulator matched to the experimental demonstrator, extensive tolerancing has shown that network parameters such as learning rate and the slope of the nonlinear neuron response function, and the nonlinearity, symmetry, and bounded nature of the conductance response are critical to achieving high performance in an NVM-based neural network [4].

Our results show that all NVM-based neural networks (not just those based on PCM) can be expected to be **highly resilient to random effects** (NVM variability, yield, and stochasticity), but will be **highly sensitive to "gradient" effects that act to steer all synaptic weights**. A learning rate just high enough to avoid network "freeze-out" is shown to be advantageous for both high accuracy and low training energy.

Simulations of ANNs with both PCM and non-filamentary bipolar RRAM based on $Pr_{1-x}Ca_xMnO_3$ (PCMO) were also discussed. In contrast to the smooth, slightly nonlinear partial-SET and asymmetric, abrupt RESET behavior of PCM, PCMO offers continuous conductance change in both directions, but exhibits significant nonlinearities (degree of conductance change depends strongly on absolute conductance). The quantitative impacts of these issues on ANN performance (classification accuracy) were discussed.

We also showed that a bidirectional NVM with a symmetric, linear conductance response of high dynamic range (each conductance step is relatively small) would be fully capable of delivering the same high classification accuracies on the MNIST handwriting digit database as a conventional, software-based implementation, ranging from >94 % when trained on 5000 examples to >97 % when trained on the full set of 60,000 training examples.

References

1. https://en.wikipedia.org/wiki/List_of_Nvidia_graphics_processing_units#Tesla (2015)
2. Breitwisch, M., Nirschl, T., Chen, C.F., Zhu, Y., Lee, M.H., Lamorey, M., Burr, G.W., Joseph, E., Schrott, A., Philipp, J.B., Cheek, R., Happ, T.D., Chen, S.H., Zaidi, S., Flaitz, P., Bruley, J., Dasaka, R., Rajendran, B., Rossnagel, S., Yang, M., Chen, Y.C., Bergmann, R., Lung, H.L., Lam, C.: Novel lithography–independent pore phase change memory. In: Symposium on VLSI Technology, pp. 100–101 (2007)
3. Burr, G.W., Narayanan, P., Shelby, R.M., Sidler, S., Boybat, I., di Nolfo, C., Leblebici, Y.: Large–scale neural networks implemented with nonvolatile memory as the synaptic weight element: comparative performance analysis (accuracy, speed, and power). In: IEDM Technical Digest, p. 4.4 (2015)
4. Burr, G.W., Shelby, R.M., di Nolfo, C., Jang, J., Shenoy, R., Narayanan, P., Virwani, K., Giacometti, E., Kurdi, B., Hwang, H.: Experimental demonstration and tolerancing of a large–scale neural network (165,000 synapses), using phase–change memory as the synaptic weight element. In: IEDM Technical Digest, p. 29.5 (2014)

5. Burr, G.W., Shelby, R.M., Sidler, S., di Nolfo, C., Jang, J., Boybat, I., Shenoy, R.S., Narayanan, P., Virwani, K., Giacometti, E.U., Kurdi, B., Hwang, H.: Experimental demonstration and tolerancing of a large-scale neural network (165,000 synapses), using phase-change memory as the synaptic weight element. IEEE Trans. Electr. Devices **62**(11), 3498–3507 (2015)
6. Gupta, S., Kaldewey, T.: Private communication (2015)
7. Hinton, G.E., Salakhutdinov, R.R.: Reducing the dimensionality of data with neural networks. Science **313**(5786), 504 (2006)
8. Jackson, B.L., Rajendran, B., Corrado, G.S., Breitwisch, M., Burr, G.W., Cheek, R., Gopalakrishnan, K., Raoux, S., Rettner, C.T., Padilla, A., Schrott, A.G., Shenoy, R.S., Kurdi, B.N., Lam, C.H., Modha, D.S.: Nanoscale electronic synapses using phase change devices. ACM J. Emerg. Technol. Comput. Syst. **9**(2), 12 (2013)
9. Jang, J.W., Park, S., Burr, G.W., Hwang, H., Jeong, Y.H.: Optimization of conductance change in $Pr_{1-x}Ca_xMnO_3$-based synaptic devices for neuromorphic systems. IEEE Electr. Device Lett. **36**(5), 457–459 (2015)
10. Jo, S.H., Chang, T., Ebong, I., Bhadviya, B.B., Mazumder, P., Lu, W.: Nanoscale memristor device as synapse in neuromorphic systems. Nano Lett. **10**(4), 1297–1301 (2010)
11. LeCun, Y., Bottou, L., Bengio, Y., Haffner, P.: Gradient-based learning applied to document recognition. Proc. IEEE **86**(11), 2278 (1998)
12. Lee, J., Jo, M., Seong, D.J., Shin, J., Hwang, H.: Materials and process aspect of cross-point RRAM (invited). Microelectron. Eng. **88**(7), 1113–1118 (2011)
13. Park, S., Kim, H., Choo, M., Noh, J., Sheri, A., Jung, S., Seo, K., Park, J., Kim, S., Lee, W., Shin, J., Lee, D., Choi, G., Woo, J., Cha, E., Jang, J., Park, C., Jeon, M., Lee, B., Lee, B., Hwang, H.: RRAM–based synapse for neuromorphic system with pattern recognition function. In: IEDM Technical Digest, p. 10.2 (2012)
14. Rajendran, B., Liu, Y., Seo, J.S., Gopalakrishnan, K., Chang, L., Friedman, D.J., Ritter, M.B.: Specifications of nanoscale devices and circuits for neuromorphic computational systems. IEEE Trans. Electr. Devices **60**(1), 246–253 (2013)
15. Rumelhart, D., Hinton, G.E., McClelland, J.L.: A general framework for parallel distributed processing. In: Parallel Distributed Processing. MIT Press (1986)
16. Suri, M., Bichler, O., Querlioz, D., Cueto, O., Perniola, L., Sousa, V., Vuillaume, D., Gamrat, C., DeSalvo, B.: Phase change memory as synapse for ultra–dense neuromorphic systems: application to complex visual pattern extraction. In: IEDM Technical Digest, p. 4.4 (2011)
17. Suri, M., Bichler, O., Querlioz, D., Palma, G., Vianello, E., Vuillaume, D., Gamrat, C., DeSalvo, B.: CBRAM devices as binary synapses for low–power stochastic neuromorphic systems: auditory (cochlea) and visual (retina) cognitive processing applications. In: IEDM Technical Digest, p. 10.3 (2012)
18. Yu, S., Gao, B., Fang, Z., Yu, H., Kang, J., Wong, H.S.P.: A neuromorphic visual system using RRAM synaptic devices with sub–pj energy and tolerance to variability: experimental characterization and large–scale modeling. In: IEDM Technical Digest, p. 10.4 (2012)

Novel Biomimetic Si Devices for Neuromorphic Computing Architecture

U. Ganguly and Bipin Rajendran

Abstract Neuromorphic computing requires low-power devices and circuits in cross-point architecture. On-chip learning is a significant challenge that requires the implementation of learning rules like spike-timing-dependent plasticity (STDP)—a method that modifies synaptic strength depending upon the time correlation between the presynaptic and postsynaptic neuron spikes in a specific function. To implement this capability in phase-change memory (PCM) or resistance RAM (RRAM)-based cross-point arrays, two schemes have been proposed in the literature where the time correlations are captured by an address event representation scheme using an universal bus or superposition of long custom waveforms. In comparison, in biology, the pulses are sharp and the time correlation information is processed at the synapse by the natural dynamics of the synapse. These are attractive attributes for minimizing power and complexity/area. Another challenge is realizing an area- and power-efficient implementation of the electronic neuron. A leaky integrate-and-fire (LIF) neuron has been implemented using analog and digital circuits which are highly power and area inefficient. To improve area and power efficiency, we have recently proposed: (i) A Si diode-based synaptic device where the charge carrier internal dynamics is used to capture the time correlation based on sharp pulses (100× sharper than custom waveforms to improve energy per spike) which can operate at 10^3–10^6 times faster than biology (providing accelerated learning) and (ii) A compact Si neuronal device that has a 60× area and 5× power benefit compared to analog implementation of neurons. These are novel devices that are based on SiGe CMOS technology, and they are highly manufacturable. The synaptic devices are based on natural transients of the impact ionization-based $n^+ p n^+$ diode (I-NPN diode). STDP and Hebbian learn-

The original version of this chapter has been revised: For detailed information please see Erratum. An erratum to this chapter can be found at DOI 10.1007/978-81-322-3703-7_11

U. Ganguly (✉)
Department of Electrical Engineering, IIT Bombay, Mumbai 400076, India
e-mail: udayan@ee.iitb.ac.in

B. Rajendran
Department of Electrical and Computer Engineering, New Jersey Institute of Technology, Newark, NJ, USA

© Springer (India) Pvt. Ltd. 2017

M. Suri (ed.), *Advances in Neuromorphic Hardware Exploiting Emerging Nanoscale Devices*, Cognitive Systems Monographs 31,
DOI 10.1007/978-81-322-3703-7_8

ing rules have been implemented. The neuron requires further modification of the I-NPN diode requiring a gating structure and some simple circuits. A leaky integrate-and-fire (LIF) neuron has also been demonstrated. Based on their device-level area and power efficiency, system-level power and area of neural networks will be highly enhanced.

1 Motivation

As von Neumann computing reaches the energy and functionality limits, neuroscience offers examples of an alternate platform of computation. Neuroscience presents two critical paradigm shifts. The first paradigm shift is the implementation of learning—which has to do with rewiring of signal paths (equivalent to modification of logic operation sequences) based on experience (equivalent to memory). In terms of hardware, this implies the integration of memory and logic at the circuit level. Such an exercise is very different from the traditional dedicated memory and logic separation (e.g., arithmetic and logic unit (ALU) from SRAM cache and DRAM). Field programmable gate arrays (FPGAs) have some level of rewiring capability. However, time evolution of rewiring at the unit level is the norm in biological systems. The second paradigm shift is clockless operation. The timing information is communicated locally by spiking neurons. This ensures a largely quiescent system that may now be extensively interconnected system without having to drive the entire system with a clock. This leads to improved overall power efficiency to system size trade-off. Thus, local memory and local time information are the basic foundations of neuromorphic systems.

2 Biological Systems, Computing Algorithms, and Electronic Hardware Equivalents

Spiking neural networks (SNN) employing time-based weight modification algorithms have the potential to realize large power-efficient learning systems. Presently, the focus has shifted to custom computer architecture and circuits on standard CMOS device platform [21]. This requires the development of an efficient electronic analog of the biological neuron and synapse supported by suitable computer architecture to enable brain-like computing. In biology, the pre-neurons are connected by synapses to the post-neurons [10]. This is mimicked by CMOS-based neurons and a resistance RAM (RRAM) as the synapse [30] as shown in Fig. 1. The strength of interconnection between pre- and post-neurons is determined by the "conductivity" of the synapse;e.g., a non-conductive synapse impedes signal transfer while a conductive synapse enables it.

A neuron may be connected to many (typically $\sim 10^4$) downstream neurons in the brain via synapses compared to a typical fan-out of 4–10 in logic circuits. The synapses may have different strengths of connectivity measured in conductance. A highly conducting synapse is said to have a high synaptic strength and vice versa. We

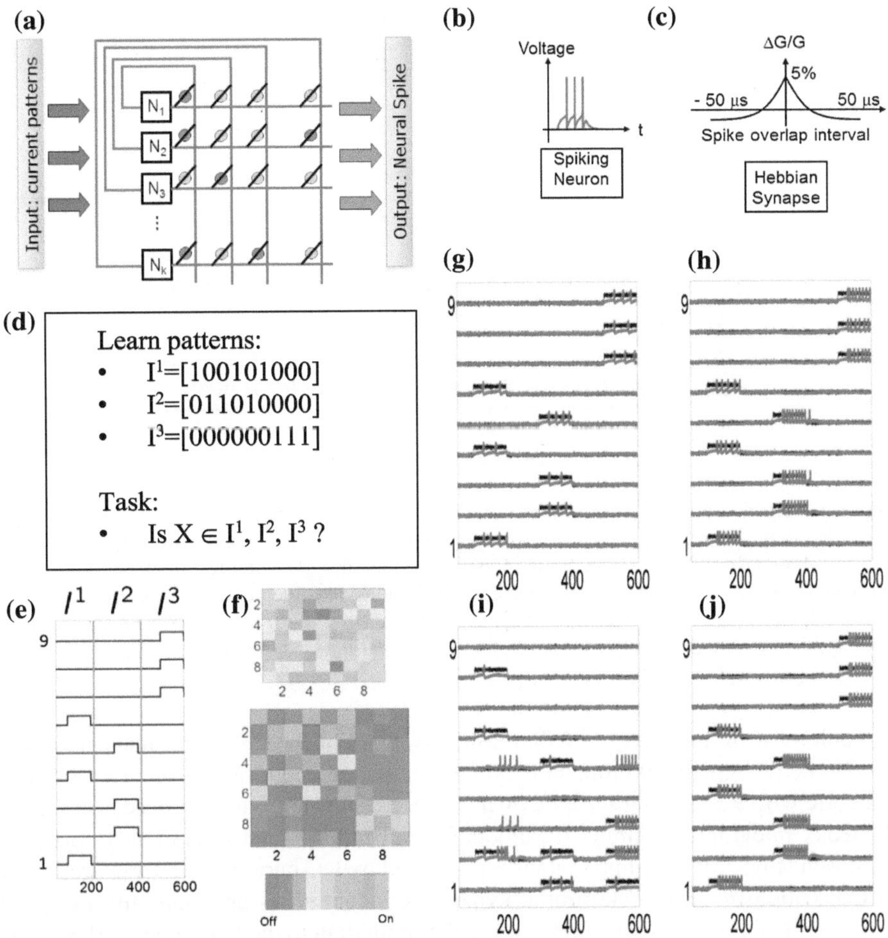

Fig. 1 **a** An array of pre- and postsynaptic neurons is connected by an n × n array of synapses based on a cross-bar array architecture. **b** A neuron with a leaky integrate-and-fire (LIF). **c** A synapse with a given STDP-based learning rule. **d** A task is provided to recognize 3 binary sequence of 9 bits using a 9 × 9 synaptic array from spurious sequences. **e** The input is represented as DC levels in time to 9 input neurons for the three training patterns. **f** Synaptic array with random weights before training and patterned weights after training. Response to inputs **g** before and **h** after training shows that after training vigorous spikes occur corresponding to input signal patterns. Test shows that **i** for "unrecognized patterns" spiking is not correlated with input **j** for "recognized patterns" spikes are highly correlated with input signal [31, 32]

present simple pattern recognition example. A cross-bar array of n × n synapses connecting n presynaptic to n postsynaptic neurons is shown. As only some synapses are conductive, it represents a circuit configuration where there is a connection between only specific pre- and postsynaptic neurons that defines a specific set of signal pathways. If the synaptic strengths of the cross-bar array are modified, the neuronal connectivity can be reconfigured. This enables a reconfigurable circuit. To give an

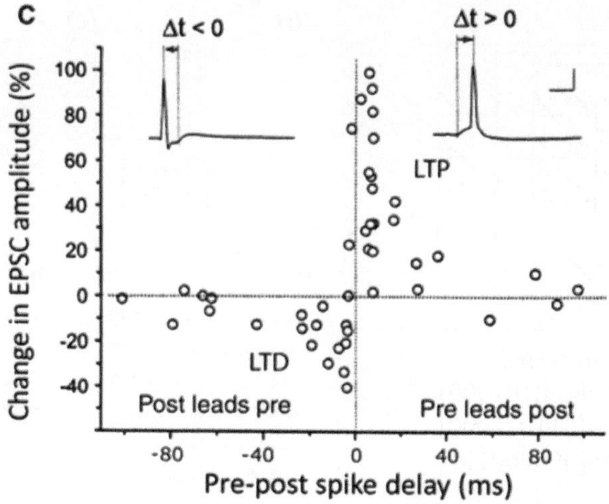

Fig. 2 Change in excitatory postsynaptic current (EPSC) which is essentially synaptic strength or conductivity as a function of pre-neuron spike time (referenced to post-neuron spike), i.e., Δt. Causal events (Δt > 0) causes increase in synaptic strength (*LTP* long-term potentiation) while anti-causal events (Δt < 0) causes a decrease (*LTD* long-term depression). A correlated event (|Δt| ∼ 0) causes strong change while uncorrelated event (|Δt| ≫ 0) causes a weak response. This response of such a nature is called spike-timing-dependent plasticity (STDP) [3, 10]

example of a simple pattern recognition task, a set of three 9-digit binary numbers need to be recognized. 9 input and output neurons are connected through a cross-bar structure with 9 × 9 synapses as shown in Fig. 2a. The neuronal response is based on the leaky integrate-and-fire (LIF) model (Fig. 2b). A specific synaptic learning rule is chosen (Fig. 2c). The input data is represented by DC current levels and presented to the 9 input neurons. The initial synaptic conductivities are randomly assigned (Fig. 2e). Training consists of presenting the input data to the neural network repeatedly. Before training, the spiking of the output neurons is not highly correlated with the input data (Fig. 2f). However, as training progresses, the spiking of the output neurons becomes vigorous and highly correlated with the input data (Fig. 2g). This also causes a change in the pattern of synaptic conductance map from random to a specific pattern (in this case approximately diagonally symmetric) (Fig. 2e). Upon training completion, if a spurious or "unrecognized" data is presented, we observe that the output spikes are sparse and not correlated with input data. However, if training pattern is presented, the output spikes are vigorous and strongly correlated with input. Thus, the neural network produces a distinct response (sparse and uncorrelated spikes) to spurious data as opposed to familiar data (vigorous and correlated spikes) which is equivalent to recognition. This basic strategy with further refinements can be demonstrated based on a computer model of this network and its evolution. These computer models have been demonstrated to be able to implement recognition tasks such as handwriting recognition and music recognition. A conventional von Neumann architecture-based computer with separate memory and logic units requires a large data transfer rate and consequently compromises energy efficiency.

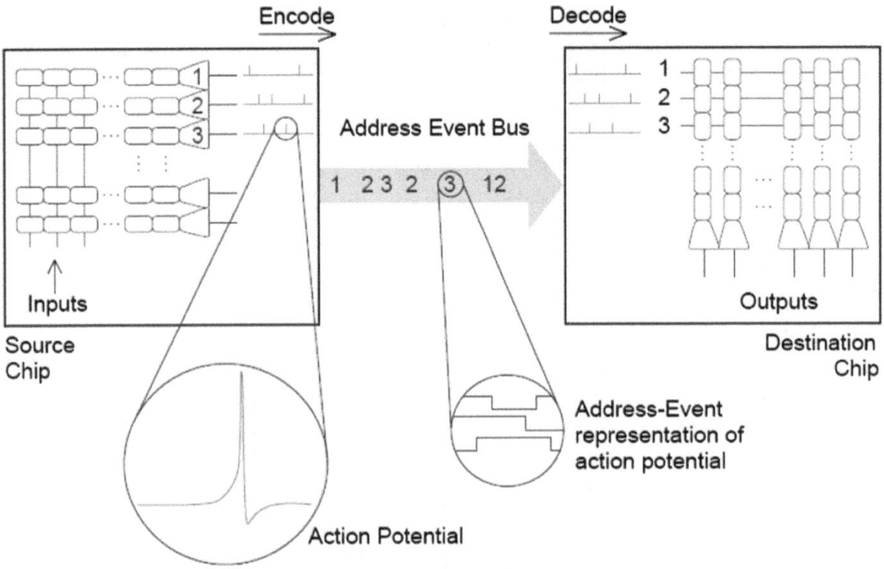

Fig. 3 Output neurons issue spikes that are encoded by address event representation (AER), send on a bus and decoded at the input of the receiving neurons [24]

A more biomimetic approach is required where both the architecture and the device mimic the brain with better fidelity. This chapter focuses on electronic devices that reproduce the functions of synapses and neurons to enable a more biomimetic architecture and further realize the advantages of the biological system (Fig. 3).

2.1 Synapse

Learning is based on the idea that a random pattern of synaptic conductivity may be trained toward a specific pattern by repeated application of a signal pattern that needs to be "memorized." The conductivity of a synapse is modified as a function of Δt, i.e., time of pre-neuron firing with post-neuron firing time as a reference. Such a Δt is essentially a measure of correlation between pre- and post-neuron firing. Figure 4 shows a biological synapse whose strength is modified by the Δt. Causal spikes ($\Delta t > 0$) increase synaptic strength while anti-causal spikes ($\Delta t < 0$) decreases it. Highly correlated spikes ($|\Delta t| \sim 0$) elicit a strong response unlike uncorrelated events ($|\Delta t| \gg 0$). Thus, conductivity of a synapse may be modified by Δt dependence of the following sort (Fig. 4) termed as spike-timing-dependent plasticity (STDP).

For implementing synapses in electronic hardware, SRAM circuits could be used, though it has significant density challenges [33]. To improve density, advanced memory devices like phase-change memory (PCM) have been pursued for system-level demonstration [16, 17, 36]. For lower power consumption, bipolar RRAM devices

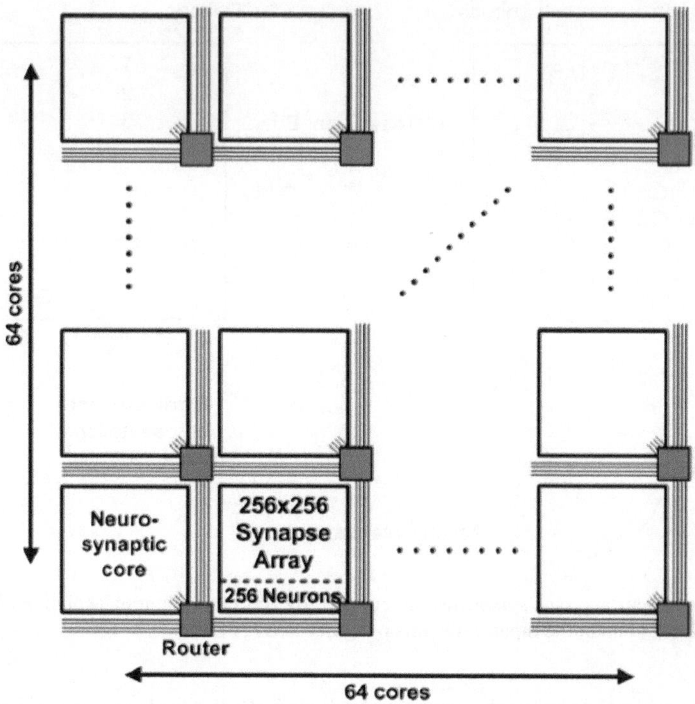

Fig. 4 High-level architecture for learning systems showing a tiled array of neuro-synaptic cores that communicate to each other using a packet routing digital mesh network. To realize a system with 1 million neurons, 4096 cores (with 256 neurons) are tiled in a 64 × 64 array [32]

have been explored [5, 12, 29, 30, 38]. RRAM provides the ability to modify conductance (ΔG) based on voltage applied (V).

The function of conversion of Δt to a conductance change (ΔG) is the central challenge [10, 31, 32]. Various memory systems provide analog G versus V characteristics as discussed above. To achieve ΔG (Δt), a two-step process is needed, i.e., first implement a timing-dependent voltage signal that can be applied to the device, V(Δt) and then use a resistance-based memory whose conductance depends on the applied voltage, G(V). To realize V(Δt), two proposals have been explored. The first one is STDP implemented by address event representation (AER)-based communication protocol. The second one is based on waveform superposition-based STDP.

2.1.1 AER-Based Communication Scheme

Neurons issue spikes in an asynchronous way. The Δt between spikes of different pre- and post-neuron combinations must be calculated first. Based on the calculated

Δt for a specific pre- and post-neuron combination, a proportional modulation of the connecting synapse is done. The calculation of Δt for various pre- and post-neuron combinations is done using address event representation (AER). Various implementation schemes and their comparison have been described in the literature [14]. A simplified description is as follows. All neurons are connected to a common bus. When a neuron spikes, it sends out its address into the bus. As the bus is fast compared to the spiking rate, the address is received all over the network "instantaneously." The receipt time of the address represents the time of spike. This way all the neurons are aware of the spiking of a specific neuron by its address and the time of receipt of the address. It now needs two further sets of information: first, its own last spike time to calculate the Δt, and second, whether the spiking neuron is connected to it. For this, a look-up table may be used. The bus may be managed by protocols of (a) address encoder and (b) address decoder. Further, a protocol to resolve concurrently occurring spikes is performed by (c) an arbiter circuit that ensures that no spike is lost even if they arrive concurrently.

There are two costs in using such a bus. First, a circuit enabling the communication protocol is an overhead. Second, the single bus presents a bandwidth constraint on total spiking rate of the system and hence the learning rate [7].

2.1.2 Waveform Superposition

An interesting observation is that some neurons are grouped together such that they communicate strongly within the group. Neuronal groups dedicated to sensory coding, e.g., retinal and cochlear neurons spike in close temporal and physical proximity while the larger population remains idle [11]. Intergroup communication is less frequent. To implement local grouping, a cross-bar array is chosen to create a neuro-synaptic core. Many such cores are tiled. Each such core is connected to other cores through a bus as described above or by a router system as shown in Fig. 4. The crossbar within the neuro-synaptic core can be driven by a simple scheme that does not require AER-based communication. The scheme involves custom waveforms that are applied from the pre- and the post-neurons when they spike.

The synapse could be implemented by a memristive RRAM device, along with a bipolar diode for nonlinearity as shown in Fig. 5a. An exemplary waveform consists of a sharp peak followed by an extended peak in the opposite polarity using analog RC delay-type circuits [31, 32]. These waveforms are applied to the two terminals of the synapse by the pre- and postsynaptic neurons when they spike. The superposition of these waveforms staggered by Δt, i.e., the spike time difference between pre- and postsynaptic neurons, creates a Δt dependent V-peak across the synaptic device (Fig. 5b). Thus, the conductivity of the RRAM is modified. As the ΔG depends upon Δt, STDP is implemented. Akin to synaptic recording [10], applying these pulses in random, our group has demonstrated STDP and Hebbian learning (Fig. 5c, d) using $Pr_{0.7}Ca_{0.3}MnO_3$-based RRAM [28] (Fig. 6).

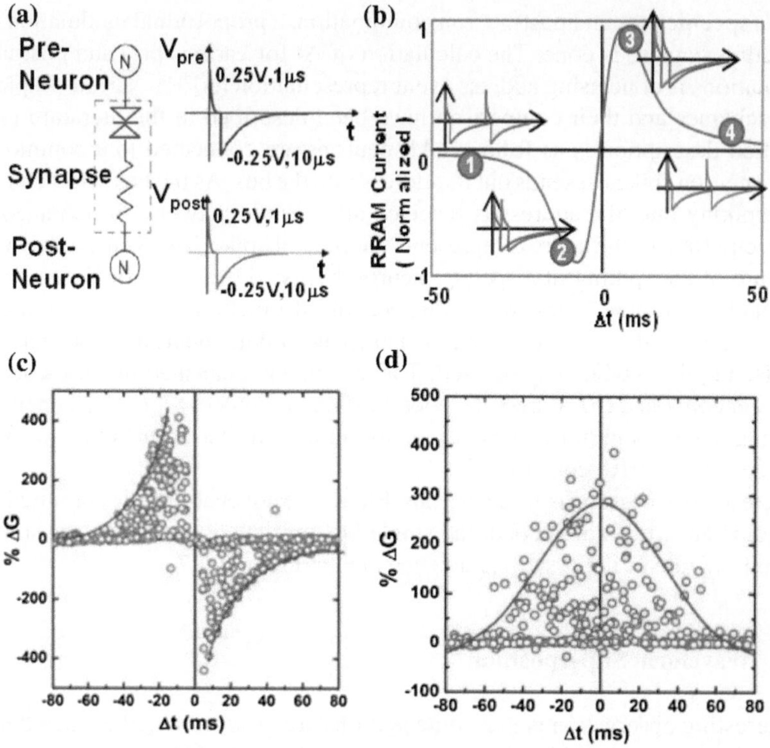

Fig. 5 **a** The synapse is represented by a memristive RRAM device along with a nonlinear bipolar diode. Custom pulses are provided by the pre- (*blue pulse*) and post(*red pulse*)-neurons. **b** The overlap of the pulses dependent on time produces different magnitudes of peak voltage across the RRAM, and hence, RRAM conductance changes. Experimental observation of PCMO (Pr0.7Mn0.3CaO3)-based synaptic plasticity showing **c** STDP and **d** Hebbian learning by random generation of custom pulses [28]

2.1.3 Motivation for Synaptic Time Keeping

As shown in Fig. 5, the neuronal pulses are long (15 μs), a significant fraction of the time correlation range (50 μs),and hence,energy-expensive. In comparison, biological pulses are sharp (∼1–5 ms) and 40–100× smaller than the time-correlation range (∼80–100 ms). In biology, such time correlations are implemented in the synapse. It has been demonstrated that this can be achieved by using the intrinsic transient behavior of a novel $4F^2$ I-NPN device [8, 9]. The device implements spike-timing-dependent resistance changes using very short and simple pulse signals similar to those issued by biological neurons [22, 23]. In the next section, we discuss the physics of this device, and also the experimentally validated device model that can be used to study and benchmark the efficiency of such neuromorphic computing architectures.

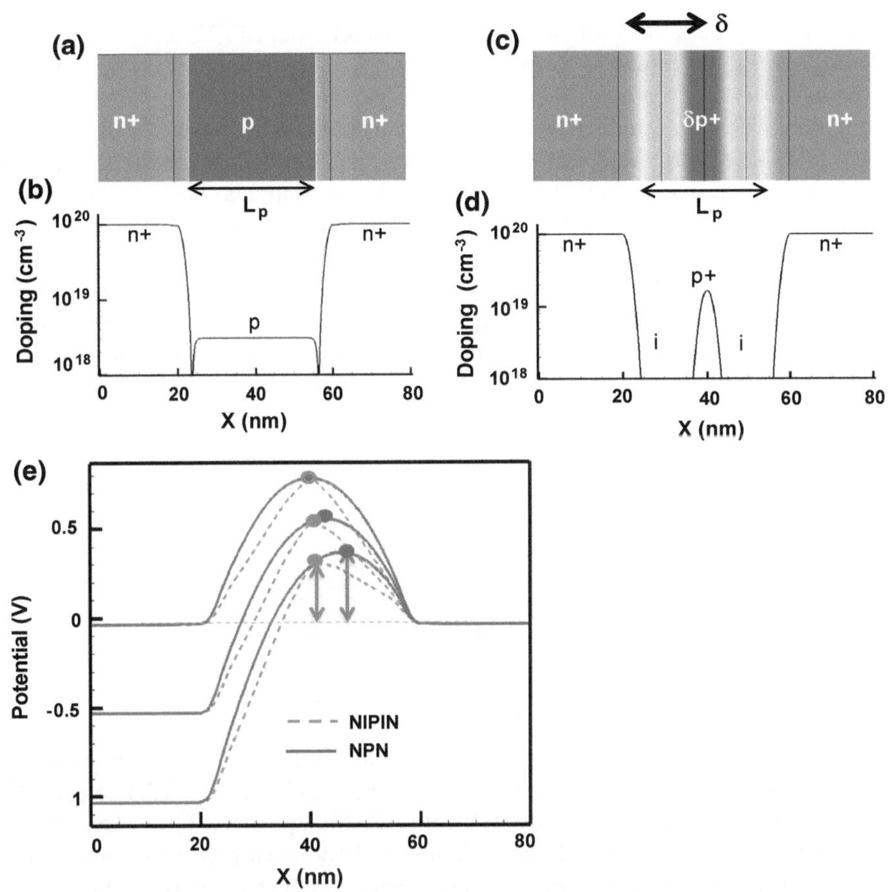

Fig. 6 A punch-through diode with **a** uniformly doped p-region in a n+ p n+ stack or NPN diode, **b** doping profile, **c** delta-doped p-region in a n+ i δp i n+ stack or NIPIN diode, **d** doping profile, **e** Applied bias dependence of peak barrier position and magnitude shows that triangular barrier (due to δ-doping) has stronger voltage modulation than parabolic barrier (uniform p-doping) [22]

2.1.4 Background of Si-Based Punch-Through Diode

The synaptic time-keeping device evolved out of a Si bidirectional selector diode. The n+/p/n+ vertical diode by epitaxial Si has been demonstrated experimentally [2, 34]. Its design space has been evaluated to demonstrate voltage scalability, nonlinearity, and asymmetric operation [18, 19]. To enable backend compatibility, low-temperature Si junctions (sub-430 °C) have been demonstrated in Si [20]. Further, low-temperature Ge epitaxy has been demonstrated from Ge-based triangular barrier selectors (Srinivasan et al. 2015—not online yet-add citation). The vertical NPN diode consists of an n+ p n+ stack with a uniformly doped p-region which is grown by Si epitaxy [4]. To improve ideality, a triangular barrier is implemented by an n+/i/δ p/i/n+ stack by Si epitaxy where the p-region is delta-doped [22, 23]. The ideality (η) is given by the extent of barrier reduction (V_b) due to applied voltage (V_a) as given by Eq. (1). Given that the p-region is in punch-through, the net charges, i.e., the depletion charges remain bias-independent in the low-bias regime. So the applied bias causes an electric field that causes barrier reduction whose extent depends on the peak position like a simple voltage divider. In the triangular barrier, the position of the electron barrier does not change. If there is any asymmetry in the position of the δ-doped p-region, it is captured in δ as shown in Fig. 5b. The ideality for bias applied to left terminal (η_L) versus right terminal (η_R) is given by equation (2). If the $\delta = L/2$ (symmetric), ideality $\eta_L = \eta_R = 2$. If $\delta \neq L/2$ (asymmetric), then ideality $\eta_L \neq \eta_R$. The average ideality is given by Eq. (3) in which the minimum $\eta_{avg} = 2$ when $\delta = L/2$.

$$\eta = \frac{V_a}{V_b} \tag{1}$$

$$\eta_L = \frac{L}{\delta}; \quad \eta_R = \frac{L}{L-\delta} \tag{2}$$

$$\eta_{avg} = \frac{\eta_L + \eta_R}{2} = \frac{L^2}{2(L-\delta)\delta} \tag{3}$$

$$\min(\eta_{avg}) = 2; \text{ when } \delta = \mathrm{L}/2 \tag{4}$$

Thus, the NPN selector provides a bidirectional nonlinear element with excellent ideality or nonlinearity. Impact ionization may be enabled in the NPN selector (I-NPN) to further improve the ideality and subthreshold slope to sub-60 mV/decade [9]. As shown in Fig. 7a in a NPN diode with uniformly doped p-region, at equilibrium, current is zero. At small bias (Fig. 7b), the barrier is reduced due to punch-through effect, akin to drain-induced barrier lowering in MOSFETs. Hence, the diode current increases exponentially. If the electrons can impact ionization at the positive terminal, an electron–hole pair is generated. The electron contributes to the diode current directly. However, the hole gets stuck in the p-well. The holes stored in the p-region in this manner reduce the electron barrier and enhances the electron current strongly. This increase in current enhances impact ionization, causes more hole storage, and sets up a positive feedback process. As the electron barrier reduces, the hole well depth reduces. Steady state is achieved when hole escapes either out of the well or by recombination matches hole generation by impact ionization (Fig. 7d). TCAD simulations have been used to demonstrate that upon adding impact ionization, the subthreshold slope improves. Figure 8 shows that a sub-60 mV/decade can be achieved. Further, using Si/SiGe/Si-based heterostructures, a sharper subthreshold slope may be enabled. The experimental evidence of low-voltage (sub-0.5 V) impact ionization has also been presented [8]. Recall that in the thermal excitation over the barrier, the minimum ideality is $\eta_{avg} = 2$. The temperature-dependent IV of NIPIN

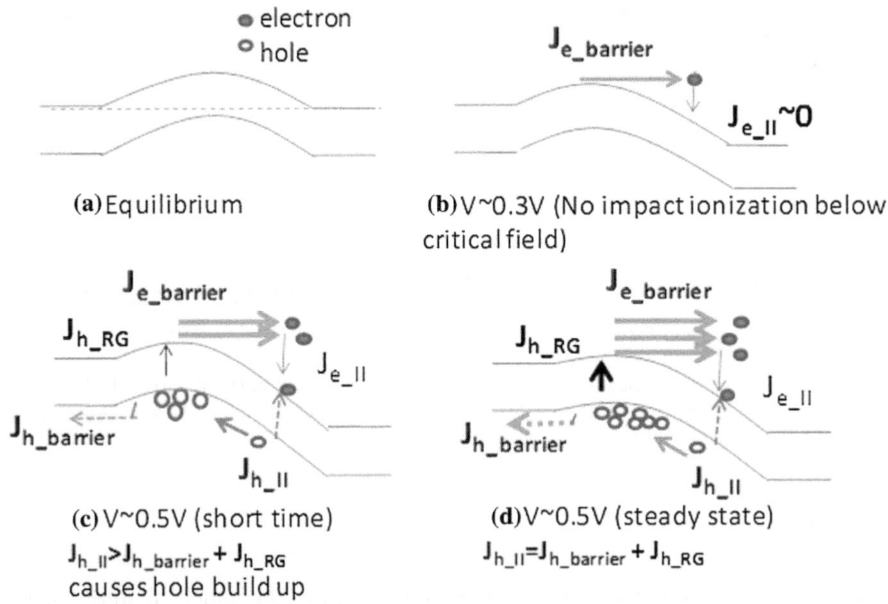

Fig. 7 Band diagram showing **a** equilibrium, **b** current due to electron conduction over the barrier, **c** beginning of impact ionization that creates hole trapping in the p-type well. This provides electrostatic barrier lower to enable more current. Higher current produces more impact ionization. A positive feedback process ensues. **d** Steady state when holes generated by impact ionization is lost by hole loss from the p-well by recombination or escape over the hole barrier [9]

Fig. 8 IV characteristics showing steep subthreshold slope of better than the thermal limit of 60 mV/decade to indicate the effect of impact ionization. Drift diffusion simulation verified by Monte Carlo [9]

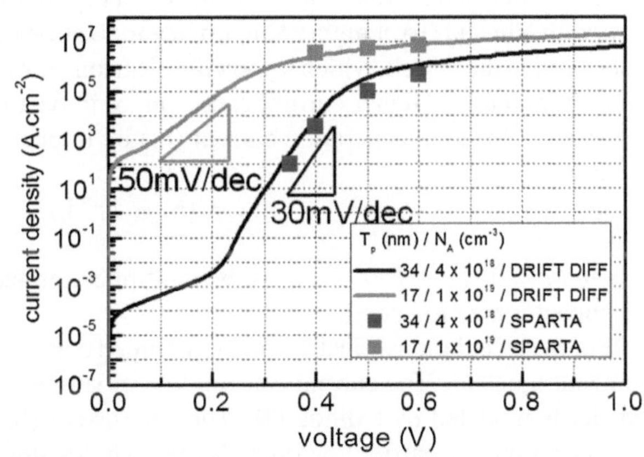

device shows that at T > 250 K, $\eta_{avg} < 2$ is observed. The voltage of η extraction is less than 0.5 V, which indicates low-voltage impact ionization. This is consistent with sub-bandgap voltage impact ionization that has been observed earlier [1].

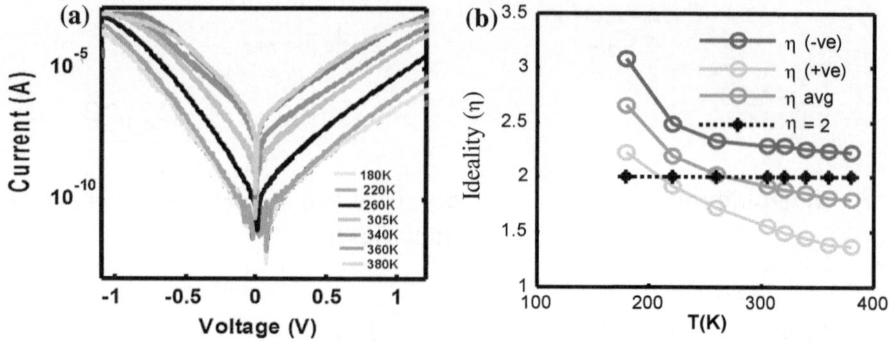

Fig. 9 **a** Temperature-dependent IV of NIPIN diode. **b** Extracted at 0.5 V versus T shows that $\eta_{avg} < 2$ at $T > 250$ K. This demonstrates that thermal excitation over the barrier alone cannot account for improved ideality. Impact ionization is responsible [8]

2.1.5 I-NPN Circuit Model

Now we discuss a compact circuit model that captures the salient features of the I-NPN device [27]. As we have seen above, the I-NPN diode has two regimes: (i) At small applied bias (V_{in}), the barrier (Φ_B) is lowered by voltage division to get exponential current as shown in Fig. 10a. This is implemented by a voltage divider circuit in Fig. 10b, where upon voltage division, the resultant voltage V_a is applied to the gate of the MOSFET operating in subthreshold region (which is essentially a voltage controlled current source) as shown in Fig. 10 to get exponential drain current (I_D) with Vin and (ii) at higher Vin, impact ionization occurs and the generated holes get stored in the p-well, which lowers the electron barrier as shown in Fig. 10c; hence, current increases. Higher current causes more impact ionization to create a positive feedback. This hole generation current (I_{II}) by impact ionization is implemented by a current source,

$$I_{II} = (M - 1) * I_D \tag{5}$$

where $M = 1/(1 - \int ae^{-\frac{b}{E}}dx)$ is the multiplication factor based on a simple model of impact ionization coefficient [6].

For the calculations, electric field is assumed constant in the I-region where impact ionization occurs. The current causing loss of stored holes (I_L) over the source hole barrier is modeled by a diode D1. The net current ($I_{II} - I_L$) charges a capacitor to produce a further barrier lowering V_b due to holes stored in the P-region. The diode current I_L depends upon total barrier lowering, i.e., $V_G(= V_a + V_b)$ applied on the gate of the MOSFET. As it already has V_b on one end, a $-V_a$ source is added to the other end to enable the bias-dependent barrier lowering. As V_b increases, leakage current increases and multiplication factor decreases. Hence, V_b will settle to a voltage such that impact ionization current becomes equal to diode leakage current. A second diode D2 is added to ensure that without impact ionization $V_b = 0$ for low

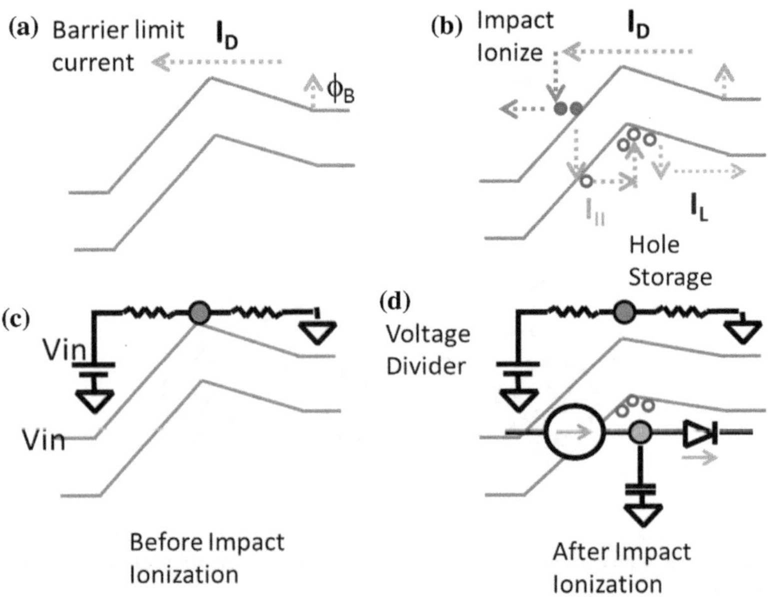

Fig. 10 **a** At low voltage, barrier lowering due to punch-through. **b** Equivalent voltage divider circuit. **c** At high voltage, impact ionization-based hole storage causes barrier reduction, current increase, and further impact ionization (positive feedback). **d** Impact ionization modeled as current source (I_{II}) and diode current (I_L) for stored hole loss. Stored holes in the capacitor produce V_b [24]

input voltages. This is physically equivalent to the drain barrier for hole leakage. Hence, a voltage source V_a is applied to the other end to enable the bias-dependent barrier increase on the drain side. Thus, the circuit elements mimic the main physical mechanisms. To obtain exponential current–voltage relationship, $V_G = V_a$ (barrier lowering by applied bias) + V_b (barrier lowering by hole storage) is applied to MOSFET M1 operating in subthreshold region. The complete equivalent circuit is shown in Fig. 11. Similar to DC I–V by TCAD simulations, I–V by SPICE shows the improved ideality upon enabling impact ionization as shown in Fig. 12a. The structure, doping, and germanium content of I-NPN device determine ideality and the threshold voltage for onset of impact ionization. The SPICE model ideality and the threshold voltage for onset of impact ionization can be engineered by selecting parameters of M, i.e., *a* and *b* as shown in Fig. 12b.

2.1.6 Transient Characteristics of I-NPN

TCAD-based turn-on transient shows that an instantaneous modulation in the turn-on and turn-off transients, respectively, is because of barrier changes due to voltage division. This is followed by a transient flat and then sharp turn-on as holes builds up

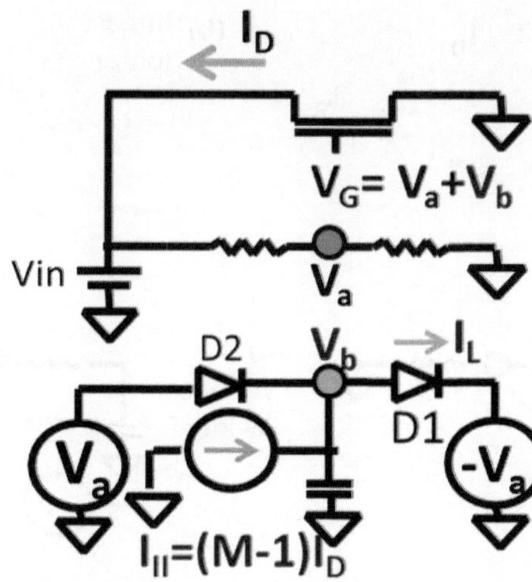

Fig. 11 $V_G = V_A + V_B$ is applied to the gate of MOSFET working in subthreshold region to obtain exponential drain current ID. Diode D2 ensures without impact ionization $V_b = 0$ for lower voltages [24]

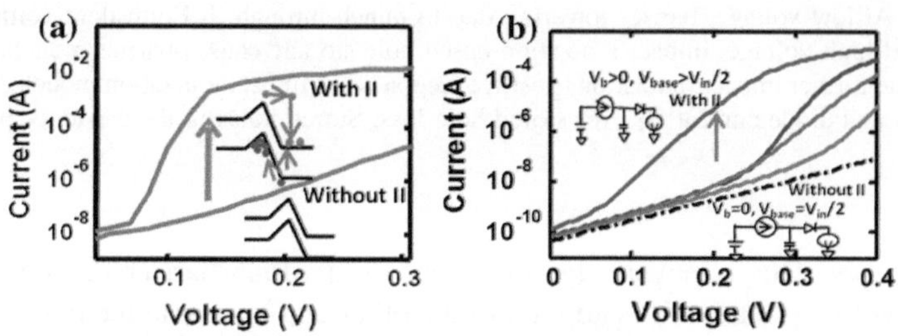

Fig. 12 **a** TCAD. **b** SPICE I–V characteristics of the device with and without II. TCAD shows ideality improvement with hole storage. SPICE shows ideality improvement with C charge up. Modifying multiplication factor M can engineer ideality [24]

with positive feedback (Fig. 8a) while turn-off shows an exponential decay timescale. The SPICE-based turn-on and turn-off transients show similar behavior (Fig. 8b). The turn-on causes the charge up of the capacitor due to the net current between I_{II} and I_L while turn-off transient is due to capacitor discharge through the D1 due to I_L. For fixed impact ionization parameters *a* and *b*, leakage diode saturation current (I_s) and capacitor C1 can be used for engineering transient properties (Figs. 13 and 14).

2.1.7 Pulse-Train Response

The TCAD-based pulse-train response shows gradual increase of I_D as holes build up during spikes and then partially leak out during the next low-bias part [4].

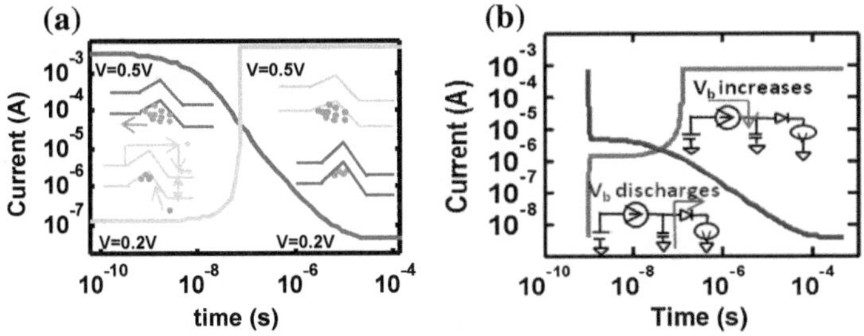

Fig. 13 Simulated turn-on ($V_{OFF} \rightarrow V_{ON}$: *red*) and turn-off ($V_{ON} \rightarrow V_{OFF}$: *blue*) transients of the device which depend upon stored hole buildup and stored hole escape in the p-region well, respectively. **a** TCAD. **b** SPICE shows charge discharge of capacitor [24]

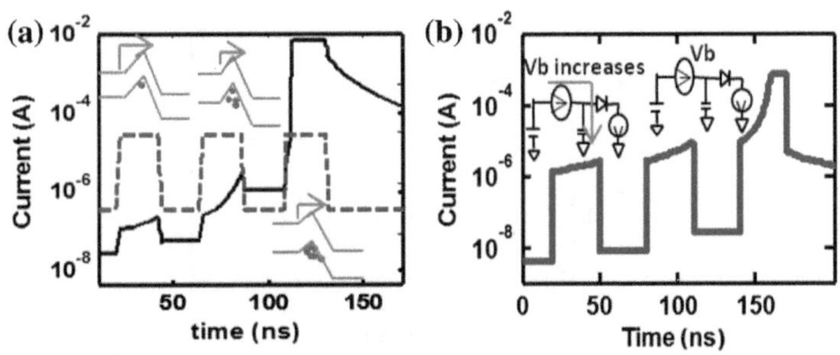

Fig. 14 The response to a pulse train shows increase in stored hole density, and the resultant ID increases by barrier reduction in **a** TCAD. **b** SPICE [24]

The SPICE response shows similar gradual current response builds up to a pulse train (Fig. 9). This demonstrates that the SPICE circuit model is able to capture the salient electrical behavior exhibited by TCAD simulations.

2.1.8 Spike-Timing-Dependent Plasticity

The TCAD simulations show that when the spike time difference is small (time difference < t_{off}), the stored holes do not get enough time to escape leading to a large I_D response for the second pulse as shown in Fig. 15. As the spike time difference increases, the stored holes are reduced and the I_D response to second pulse decreases. The SPICE simulations show the same behavior as it is able to charge up the capacitor during spiking and partially discharge the capacitor based on the time difference. This is the synaptic time-keeping function. Using this time-keeping function in series with RRAM device, time difference can be converted to conductance change and can then generate STDP and other timing-dependent plasticity behaviors. Based on these features, an STDP learning rule has been demonstrated in SPICE and compared to the TCAD results as shown in Fig. 16.

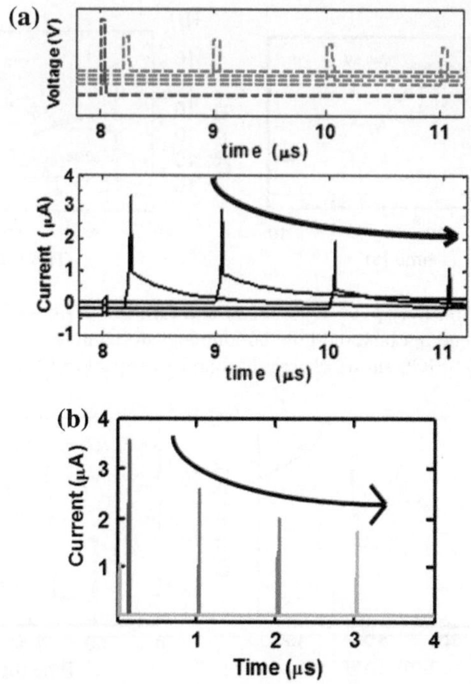

Fig. 15 Effect of spike time difference shows high current for low time difference between pulses. **a** TCAD—1st graph corresponds to two input pulses given with different time differences; 2nd graph shows output current for different sets of pulses. **b** SPICE model: output current for two pulses with increasing time difference [24]

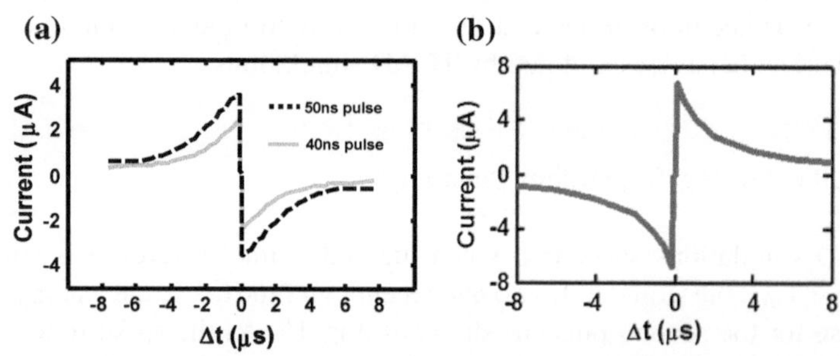

Fig. 16 Hebbian learning can be implemented in **a** TCAD and **b** SPICE. Trends are identical [24]

2.1.9 Advantages of Implementing Proposed STDP

The small ($4F^2$) footprint enables high-density cross-point arrays. It eliminates large waveform generation circuits at the periphery of the synaptic array which is 7× larger than the array itself—leading to proportionally higher areal density. The speed of learning is improved by a factor of 20–100× as long waveforms are replaced by short pulses which are approximately 20–100× smaller than correlation time range. Choice with a fast RRAM (μs set/reset) may provide a further acceleration from biological pulse timescales (~ms) by a factor of 1000×. Finally, unlike neuronal implementation, the synaptic implementation of time correlation is truly biomimetic with a large ratio between spike time (pulse width) and learning timescale.

2.2 Neuron

While the synapse is responsible for reconfiguring the network, the neurons essentially form the nodes of the network. As the neuron is formed of many complex biochemical systems, a simple mathematical models have been developed to emulate the behavior [13, 15]. The simplest model is the leaky integrate-and-fire neuron model [35] which may be described as follows. A neuron receives signals as current spikes (I_{in}) via synapses from many pre-neurons shown in Fig. 17a. The simple leaky integrate-and-fire (LIF) model of the neuron describes that the neuron will integrate the I_{in} (akin to charging of a capacitor) and partly discharge between the I_{in} spikes (as

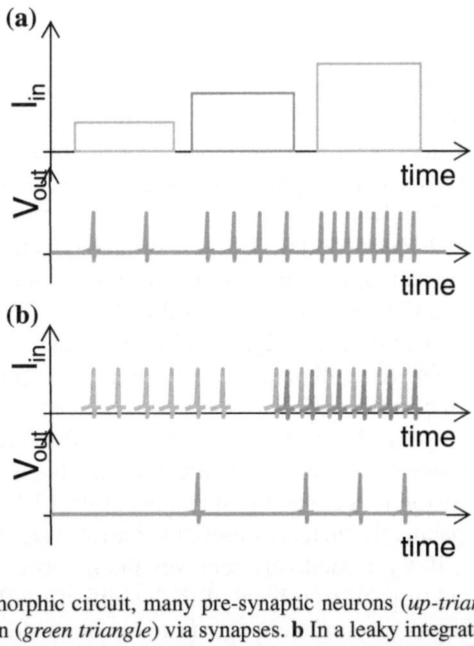

Fig. 17 **a** In a neuromorphic circuit, many pre-synaptic neurons (*up-triangles*) are connected to the postsynaptic neuron (*green triangle*) via synapses. **b** In a leaky integrate-and-fire (LIF) neuron model, as the pre-neuron issues current spikes, the post neuron integrates (I) the incoming current, but the accumulated charge can leak away (L) in between spikes. When the potential across the capacitor reaches a threshold level, the post-neuron spikes and the resets (all charge is lost) are to start all over again. **c** As the input-current level increases, threshold is reached faster and spikes are more frequent [27]

if the capacitor is leaky). When the voltage across the capacitor exceeds a threshold level, the neuron will issue a spike (fire) and then reset (lose all integrated charge) to start the process again (Fig. 1b). Thus, as the magnitude of input current I_{in} increases, the neuron spiking frequency will increase as shown in Fig. 17c as integrating a higher input current results in reaching threshold levels in shorter timescales.

2.2.1 Digital and Analog Circuit-Based Neuron

Various analog and digital implementations of the LIF neurons have been presented [31–33]. The main idea has been a circuit implementation based on standard CMOS devices. However, each approach has a different power, and area performance based on the number of transistors required to implement the neuronal function [31–33]. Thus, a compact implementation with fewer circuit elements may be more advantageous from an area and power perspective.

2.2.2 A Gated I-NPN Based Si Neuron

Low-voltage impact ionization (II)-based n+/p/n+ selector (I-NPN) has been proposed [9, 22], and the physics has been experimentally validated [25] including impact ionization at sub-0.5 V in Si NPN device as seen earlier [8]. It has been shown that such devices could be also adapted to mimic neuronal integration function [26]. An exemplary n+ (10^{20} cm^3, Si, 20 nm)/p (4×10^{18} cm^3, $Si_{0.9}Ge_{0.1}$, 35 nm)/n + (10^{20} cm^3, Si, 20 nm) device on a 20 nm node SOI is shown in Fig. 18a. During turn-on (0.2 V → 0.5 V), electrons undergo II to generate electron (e)–hole (h) pair as shown in Fig. 18b. The *hole is stored* in the p-well to reduce the e-barrier and enhance e-current which causes further II and h-storage (positive feedback). Steady state occurs when h-current (J_o) due to h-escape from well cancels out h-generation current (J_{in}) as shown in Fig. 18c. During turn-off (0.5 V → 0.2 V), instant e-barrier increase occurs due to electrostatics (Fig. 18d). Then, further barrier increase occurs as stored holes recombine to equilibrium slowly (Fig. 18e). Thus, I-NPN is able to integrate and fire during turn-on. However, neuronal fast reset requires fast loss of stored holes after the voltage has peaked during firing. This necessitates a voltage-dependent control of stored h-escape from the p-well. This is obtained by adding a gated p-region extension to a contact to the conventional I-NPN (Fig. 18f). If the gate voltage (V_g) capacitively increases h-escape barrier (Fig. 18h), then h-storage may occur. However, if V_g capacitively removes the h-barrier, fast hole loss may occur. The turn-on and turn-off transients of the modified I-NPN device show that $V_g = 0.8$ V is able to provide sufficient barrier to enable h-storage during turn on. During turn-off, $V_g = 0.8$ V leads to slow h-loss (Fig. 19). However, at $V_g = 0$ V, the h-loss is fast. Thus, a V_g-controlled h-loss capability is added. Finally, to ensure that reset (i.e., h-loss) occurs right after spiking, a resistance R1 is added in series with modified I-NPN as shown in Fig. 20a. When I-NPN is off, V_o is low and transistor M is off, so V_g is high. Hence, h-storage and buildup occurs. As I-NPN turns on, V_o

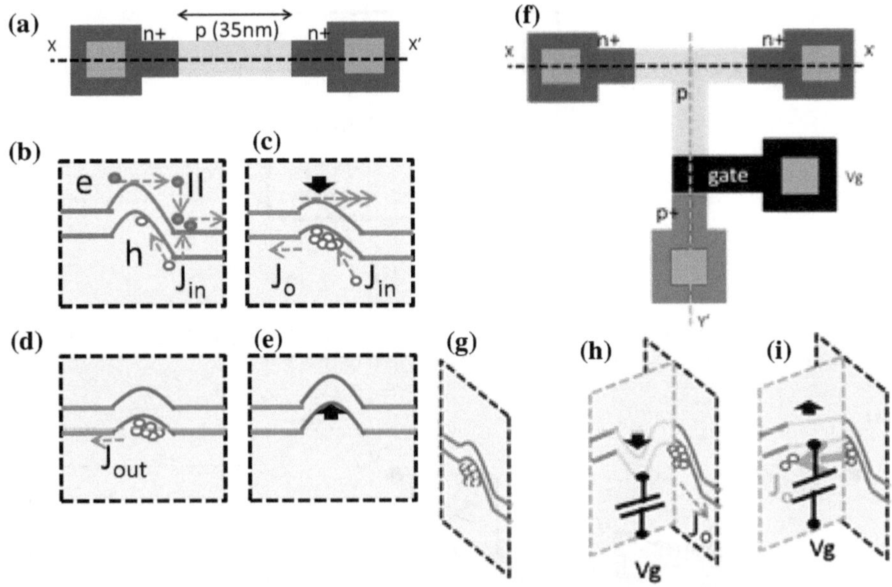

Fig. 18 **a** Structure of impact ionization-based NPN device (I-NPN). **b** During turn-on ($0 \rightarrow V$), impact ionization occurs to generate holes. Holes get stored in the p-well to reduce e-barrier increase current, producing further impact ionization. **c** Steady state is reached when hole current generation (J_{in}) is equal to hole current (J_o) escaping the well from decreasing h-well. **d** During turn-off ($V \rightarrow 0$), e-barrier is low due to stored holes. **d** Holes escape by recombination "slowly" to restore equilibrium to increase barrier. **e** To enable V-controlled hole escape rate, a MOS-based p-region in added. **f** Band diagram is the XX′ direction equivalent to band diagram in (**d**). **g** Band diagram in YY′ added to show that stored holes cannot escape if V_g capacitively produces barrier for holes (**h**) This hole barrier can be removed by V_g to enable fast charge loss. This enables V_g-controlled hole escape [27]

increases sharply (Fig. 20b) which turns the transistor (M) on, and V_g becomes low, h-barrier is reduced, and h-loss occurs. I-NPN current reduces as well. So V_o reduces and M turns off. However, to ensure that V_g remains low to ensure sufficient time for h-loss, the V_g, i.e., C_2 charge-up time ($\tau_2 = R_2C_2$) is longer than the V_o decay timescale ($\tau_1 = R_1C_1$). Thus, stored holes are fully drained to ensure a full reset. Simulations show that DC input produces spikes (Fig. 21a). The spike frequency increases with the magnitude of the DC input. In a scenario closer to operational situation, a pulse-train input causes the neuronal device to spike at regular intervals. This demonstrates the functional verification of the neuronal device based on TCAD (Fig. 22).

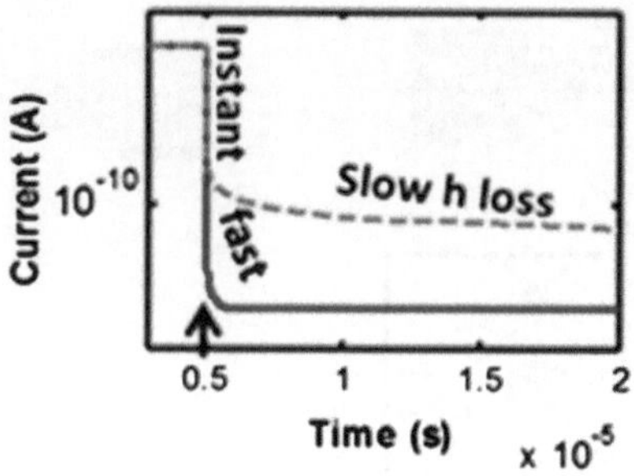

Fig. 19 IV turn-off transient with V_g off (*solid*) and on (*dashed*) showing that V_g can control hole loss rate effectively (fast vs. slow) after initial instant drop [27]

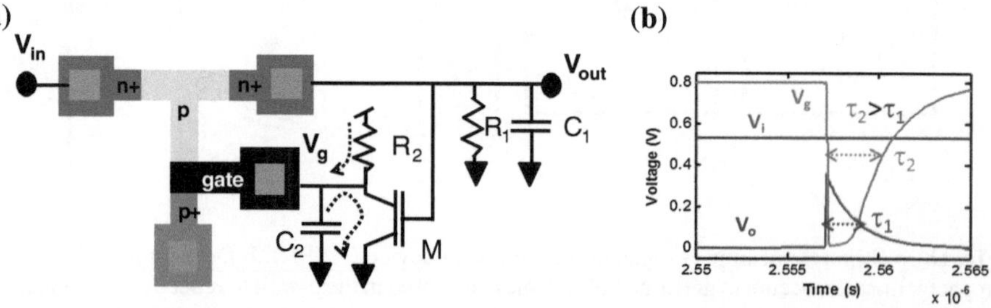

Fig. 20 **a** The complete circuit of the neuron. **b** Transient of V_{in}, V_o, and V_g. When V_{in} is low, NPN device is off, I_1 is low, and V_o is low. So M is off, hence V_g is high and I_h is low. So holes will be stored. When V_{in} turns on sharply, I_1 current increases with I-NPN turn transient timescale and V_o goes from low to high. At high V_o, $V_o > V_T$ of transistor M, M turns on to discharge C_2 quickly. Thus, V_g switches from high to low fast to increase I_h sharply and drain all the stored holes. So I_1 drops and hence V_o drops to turn off M. However, C_2 charges through R_2 slowly ($\tau = R_2C_2$) to ensure that V_g remains low for all stored holes to drain completely [27]

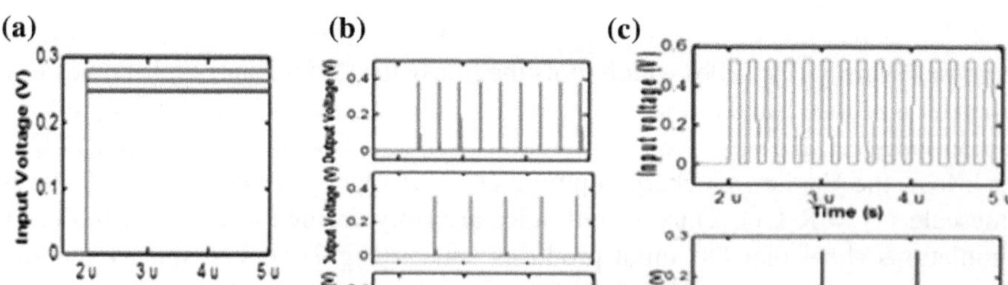

Fig. 21 **a** DC input cause. **b** Spikes in the neuron and higher DC levels produce more frequency spikes. **c** Spike train inputs produce spike output at lower frequency [27]

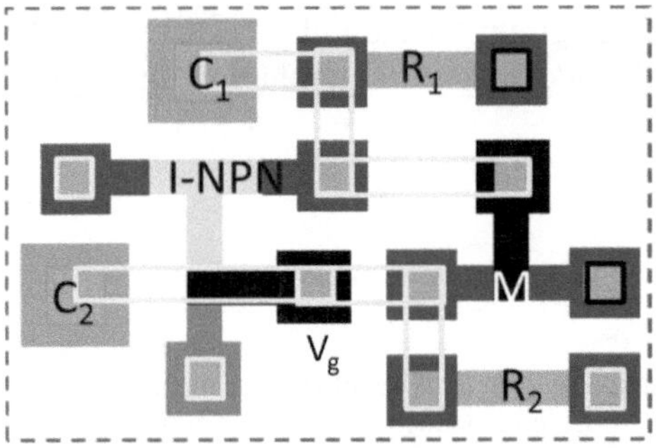

Fig. 22 Layout of the neuron to show a max 225 * F^2 size [27]

Table 1 Performance benchmark

	Analog neuron 10 nm node	Neuron device 20 nm node (this work)	Benefit
Area	5.6 μm^2	0.09 μm^2	>60×
Power	65 nW	11.5 nW	>5×

2.2.3 Area and Power

The layout of the neuron circuit shows a 225 * F^2 area assuming 20 nm technology node. Capacitors of 1fF may require a deep trench technology (capable of C_{max} = 20 fF) with a footprint of 0.013 μm^2 at 32 nm node [37]. This is approximately 14 * F^2. Power estimates showed that active power was 11.5 nW. Table 1 shows a comparison with an analog design to highlight the ultra high density (>60×) and high power efficiency (>6×) of the neuronal device compared to an analog neuron. Further, our circuit issues spikes at ~1–10 MHz, providing an acceleration factor of ~10,000 compared to biology (10 Hz).

3 Conclusions

Even though various implementations of neuromorphic circuits exist based on standard CMOS devices, a biomimetic synapse capable of fast, energy-efficient adaptation is still a challenge. We have discussed several device concepts that have been proposed and demonstrated by our groups in this chapter. We discussed a silicon synaptic time-keeping device based on I-NPN device technology that enables STDP based on simple spikes. The method enables lower power and simplification of neuronal circuits. Another critical challenge is the neuron based on standard CMOS analog or digital devices. We have shown that an I-NPN device could be used to mimic basic neuronal behavior in a highly area- and power-efficient manner. Circuit models developed in our group for these devices have also been discussed, which could be used to benchmark large neuromorphic circuits in performing various learning and recognition tasks.

References

1. Anil, K.G., Mahapatra, S., Eisele, I.: A detailed experimental investigation of impact ionization in n-channel metal-oxide-semiconductor field-effect-transistors at very low drain voltages. Solid-State Electron. **47**(6), 995–1001 (2003). doi:10.1016/S0038-1101(02)00458-6
2. Bafna, P., Karkare, P., Srinivasan, S., Chopra, S., Lashkare, S., Kim, Y., ... Ganguly, U.: Epitaxial Si punch-through based selector for bipolar RRAM. In: 2012 70th Annual Device Research Conference (DRC) (2012). doi:10.1109/DRC.2012.6256979
3. Bi, G., Poo, M.: Synaptic modifications in cultured hippocampal neurons: dependence on spike timing, synaptic strength, and postsynaptic cell type. J. Neurosci. **18**(24), 10464–10472 (1998)
4. Chopra, S., Bafna, P., Karkare, P., Srinivasan, S., Lashkare, S., Kumbhare, P., et al.: A two terminal vertical selector device for bipolar RRAM. Meeting Abstracts, No. 37, p. 2804 (2012)
5. Chou, T., Liu, J.-C., Chiu, L.-W., Wang, I.-T., Tsai, C.-M., Hou, T.-H.: Neuromorphic pattern learning using HBM electronic synapse with excitatory and inhibitory plasticity. In: 2015 International Symposium on VLSI Technology, Systems and Application (VLSI-TSA) (2015). doi:10.1109/VLSI-TSA.2015.7117582
6. Chynoweth, A.G.: Ionization rates for electrons and holes in silicon. Phys. Rev. **109**(5), 1537–1540 (1958). doi:10.1103/PhysRev.109.1537
7. Culurciello, E., Andreou, A.G.: A comparative study of access topologies for chip-level address-event communication channels. IEEE Trans. Neural Netw. (2003). doi:10.1109/TNN.2003.816385
8. Das, B., Meshram, R., Ostwal, V., Schulze, J., Ganguly, U.: Observation of impact ionization at sub-0.5V and resultant improvement in ideality in I-NPN selector device by Si epitaxy for RRAM applications. In: 2014 72nd Annual Device Research Conference (DRC) (2014). doi:10.1109/DRC.2014.6872336
9. Deshmukh, S., Lashkare, S., Rajendran, B., Ganguly, U.: I-NPN: A sub-60mV/decade, sub-0.6V selection diode for STTRAM. In: 2013 71st Annual Device Research Conference (DRC) (2013). doi:10.1109/DRC.2013.6633819
10. Feldman, D.E.: The spike-timing dependence of plasticity. Neuron **75**(4), 556–571 (2012). doi:10.1016/j.neuron.2012.08.001
11. Fried, S.I., Cai, C., Ren, Q.: High frequency electric stimulation of retinal neurons elicits physiological signaling patterns. In: Conference Proceedings?: ... Annual International Conference of the IEEE Engineering in Medicine and Biology Society. IEEE Engineering in Medicine and Biology Society. Annual Conference, 2011, pp. 1077–1080 (2011). doi:10.1109/IEMBS.2011.6090251
12. Garbin, D., Vianello, E., Bichler, O., Rafhay, Q., Gamrat, C., Ghibaudo, G., Perniola, L.: HfO_2-based OxRAM devices as synapses for convolutional neural networks. IEEE Trans. Electron Devices (2015). doi:10.1109/TED.2015.2440102
13. Hodgkin, A.L., Huxley, A.F.: A quantitative description of membrane current and its application to conduction and excitation in nerve. J. Physiol. **117**(4), 500–544 (1952)
14. Imam, N., Manohar, R.: Address-Event Communication Using Token-Ring Mutual Exclusion. In: 2011 17th IEEE International Symposium on Asynchronous Circuits and Systems (ASYNC) (2011). doi:10.1109/ASYNC.2011.20
15. Izhikevich, E.M.: Simple model of spiking neurons. IEEE Trans. Neural Netw. **14**(6), 1569–1572 (2003)

16. Jackson, B.L., Rajendran, B., Corrado, G.S., Breitwisch, M., Burr, G.W., Cheek, R., ... Modha, D.S.: Nanoscale electronic synapses using phase change devices. J. Emerg. Technol. Comput. Syst. **9**(2), 12:1–12:20 (2013). doi:10.1145/2463585.2463588
17. Kuzum, D., Jeyasingh, R.G.D., Wong, H.-S.P.: Energy efficient programming of nanoelectronic synaptic devices for large-scale implementation of associative and temporal sequence learning. In: 2011 IEEE International Electron Devices Meeting (IEDM) (2011). doi:10.1109/IEDM.2011.6131643
18. Lashkare, S., Karkare, P., Bafna, P., Deshmukh, S., Srinivasan, V.S.S., Lodha, S., Ganguly, U.: Design of epitaxial Si punch-through diode based selector for high density bipolar RRAM. In: 2012 International Conference on Emerging Electronics (ICEE) (2012). doi:10.1109/ICEmElec.2012.6636237
19. Lashkare, S., Karkare, P., Bafna, P., Raju, M.V.S., Srinivasan, V.S.S., Lodha, S., ... Chopra, S.: A bipolar RRAM selector with designable polarity dependent on-voltage asymmetry. In: 2013 5th IEEE International Memory Workshop (IMW) (2013). doi:10.1109/IMW.2013.6582128
20. Mandapati, R., Shrivastava, S., Das, B., Sushama, Ostwal, V., Schulze, J., Ganguly, U.: High performance sub-430°C epitaxial silicon PIN selector for 3D RRAM. In: 2014 72nd Annual Device Research Conference (DRC) (2014). doi:10.1109/DRC.2014.6872387
21. Merolla, P., Arthur, J., Akopyan, F., Imam, N., Manohar, R., Modha, D.S.: A digital neurosynaptic core using embedded crossbar memory with 45pJ per spike in 45nm. In: 2011 IEEE Custom Integrated Circuits Conference (CICC) (2011). doi:10.1109/CICC.2011.6055294
22. Meshram, R., Das, B., Mandapati, R., Lashkare, S., Deshmukh, S., Lodha, S., ... Schulze, J.: High performance triangular barrier engineered NIPIN selector for bipolar RRAM. In: 2014 IEEE 6th International Memory Workshop (IMW) (2014). doi:10.1109/IMW.2014.6849388
23. Meshram, R., Rajendran, B., Ganguly, U.: Biomimetic $4F^2$ synapse with intrinsic timescale for pulse based STDP by I-NPN selection device. In: 2014 72nd Annual Device Research Conference (DRC) (2014). doi:10.1109/DRC.2014.6872301
24. Mitra, S., Indiveri, G., Fusi, S.: Learning to classify complex patterns using a VLSI network of spiking neurons. In: Proceedings of the 21th Conference on Advances in Neural Information Processing Systems (NIPS) (2008). doi:10.3929/ethz-a-005665372
25. Moon, D.-I., Choi, S.-J., Kim, S., Oh, J.-S., Kim, Y.-S., Choi, Y.-K.: Vertically integrated unidirectional biristor. IEEE Electron Device Lett. **32**(11), 1483–1485 (2011). doi:10.1109/LED.2011.2163698
26. Ostwal, V., Meshram, R., Rajendran, R., Ganguly, U.: An ultra-compact and low power neuron based on SOI platform. In: VLSI TSA. Taiwan (2015) doi:10.1109/VLSI-TSA.2015.7117569
27. Ostwal, V., Rajendran, B., Ganguly, U.: A circuit model for a Si-based biomimetic synaptic timekeeping device. In: SISPAD (2015). doi:10.1109/SISPAD.2015.7292324
28. Panwar, N., Kumar, D., Upadhyay, N.K., Arya, P., Ganguly, U., Rajendran, B.: Memristive synaptic plasticity in Pr<inf>0.7</inf>Ca<inf>0.3</inf>MnO<inf>3</inf> RRAM by bio-mimetic programming. In: 2014 72nd Annual Device Research Conference (DRC) (2014). doi:10.1109/DRC.2014.6872334
29. Park, S., Kim, H., Choo, M., Noh, J., Sheri, A., Jung, S., ... Hwang, H.: RRAM-based synapse for neuromorphic system with pattern recognition function. In: 2012 IEEE International Electron Devices Meeting (IEDM) (2012). doi:10.1109/IEDM.2012.6479016
30. Park, S., Noh, J., Choo, M.-L., Sheri, A.M., Chang, M., Kim, Y.-B.: Hwang, H.: Nanoscale RRAM-based synaptic electronics: toward a neuromorphic computing device. Nanotechnology **24**(38), 384009 (2013). doi:10.1088/0957-4484/24/38/384009
31. Rajendran, B., Liu, Y., Seo, J., Gopalakrishnan, K., Chang, L., Friedman, D., Ritter, M.: RRAM devices for large neuromorphic systems. In: Non-Volatile Memories Workshop (2013)
32. Rajendran, B., Member, S., Liu, Y., Seo, J., Gopalakrishnan, K., Chang, L.: Ritter, M.B.: Specifications of nanoscale devices and circuits for neuromorphic computational systems. IEEE Trans. Electron Devices **60**(1), 246–253 (2013)
33. Seo, J., Brezzo, B., Liu, Y., Parker, B.D., Esser, S.K., Montoye, R.K., ... Friedman, D.J.: A 45nm CMOS neuromorphic chip with a scalable architecture for learning in networks of spiking neurons. In: 2011 IEEE Custom Integrated Circuits Conference (CICC) (2011). doi:10.1109/CICC.2011.6055293

34. Srinivasan, V.S.S., Chopra, S., Karkare, P., Bafna, P., Lashkare, S., Kumbhare, P., Ganguly, U.: Punchthrough-diode-based bipolar RRAM selector by Si epitaxy. IEEE Electron Device Lett. (2012). doi:10.1109/LED.2012.2209394
35. Stein, R.B.: The frequency of nerve action potentials generated by applied currents. Proc. R. Soc. London B: Biol. Sci. **167**(1006), 64–86. http://rspb.royalsocietypublishing.org/content/167/1006/64.abstract (1967)
36. Suri, M., Bichler, O., Querlioz, D., Cueto, O., Perniola, L., Sousa, V., ... DeSalvo, B.: Phase change memory as synapse for ultra-dense neuromorphic systems: application to complex visual pattern extraction. In: 2011 International Electron Devices Meeting, pp. 4.4.1–4.4.4 (2011). doi:10.1109/IEDM.2011.6131488
37. Wang, G., Cheng, K., Ho, H., Faltermeier, J., Kong, W., Kim, H., ... Iyer, S.S.: A 0.127 μm^2 high performance 65nm SOI based embedded DRAM for on-processor applications. In: 2006 International Electron Devices Meeting, vol. 1, pp. 0–3 (2006)
38. Wang, I.-T., Lin, Y.-C., Wang, Y.-F., Hsu, C.-W., Hou, T.-H.: 3D synaptic architecture with ultralow sub-10 fJ energy per spike for neuromorphic computation. In: 2014 IEEE International Electron Devices Meeting (IEDM) (2014). doi:10.1109/IEDM.2014.7047127

Exploiting Variability in Resistive Memory Devices for Cognitive Systems

Vivek Parmar and Manan Suri

Abstract In literature, different approaches point to the use of different resistive memory (RRAM) device families such as PCM [1], OxRAM, CBRAM [2], and STT-MRAM [3] for synaptic emulation in dedicated neuromorphic hardware. Most of these works justify the use of RRAM devices in hybrid learning hardware on grounds of their inherent advantages, such as ultra-high density, high endurance, high retention, CMOS compatibility, possibility of 3D integration, and low power consumption [4]. However, with the advent of more complex learning and weight update algorithms (beyond-STDP kinds), for example the ones inspired from Machine Learning, the peripheral synaptic circuit overhead considerably increases. Thus, use of RRAM cannot be justified on the merits of device properties alone. A more application-oriented approach is needed to further strengthen the case of RRAM devices in such systems that exploit the device properties also for peripheral nonsynaptic and learning circuitry, beyond the usual synaptic application alone.In this chapter, we discuss two novel designs utilizing the inherent variability in resistive memory devices to successfully implement modified versions of Extreme Learning Machines and Restricted Boltzmann Machines in hardware.

1 Introduction

In literature, several approaches point to the use of different resistive memory (RRAM) device families such as PCM [1], OxRAM, CBRAM [2], and STT-MRAM [3] for synaptic emulation in dedicated neuromorphic hardware. Most of these works justify the use of RRAM devices in hybrid learning hardware on grounds of their inherent advantages, such as ultra-high density, high endurance, high retention, CMOS compatibility, possibility of 3D integration, and low power consumption

V. Parmar · M. Suri (✉)
IIT Delhi, Hauz Khas, New Delhi 110016, India
e-mail: manansuri@ee.iitd.ac.in

V. Parmar
e-mail: vivek.p.1990@ieee.org

© Springer (India) Pvt. Ltd. 2017

M. Suri (ed.), *Advances in Neuromorphic Hardware Exploiting Emerging Nanoscale Devices*, Cognitive Systems Monographs 31,
DOI 10.1007/978-81-322-3703-7_9

[4]. However, in order to implement more complex learning and weight update algorithms (beyond-STDP kinds), for example the ones inspired from Machine Learning, the peripheral synaptic circuit overhead considerably increases. Thus, use of RRAM cannot be justified on the merits of device properties alone. A more application-oriented approach is needed to further strengthen the case of RRAM devices in such systems that exploit the device properties also for peripheral nonsynaptic and learning circuitry beyond the usual synaptic application. In this chapter, we discuss two such novel designs utilizing RRAM devices for implementation of modified Extreme Learning Machines (ELMs) and Restricted Boltzmann Machines (RBMs) in hardware. The ELM design utilizes device-to-device variability present in CBRAM and OxRAM devices to implement the random input weights. The RBM utilizes cycle-to-cycle variability in OxRAM devices to implement stochasticity in neuron activation and perform bioinspired learning. Thus, both designs tend to exploit RRAM's inherent variability as an advantage to implement learning in hardware without using RNGs (random number generators), thus effectively reducing the power and area requirements of implementation while still performing high-speed learning unlike traditional spiking neural network (SNN) circuits implemented using RRAM.

2 Nanoscale Filamentary RRAM

RRAM devices are two-terminal MIM-type structures (metal–insulator–metal) sandwiching an active insulator layer, between metallic electrodes (see Fig. 1). The active layer exhibits reversible nonvolatile switching behavior on application of appropriate programming current/voltage across the device terminals. In the case of filamentary RRAM devices, formation of a conductive filament in the active layer leads the device to a low-resistance (LRS/On) SET-state, while dissolution of the filament puts the device in a high-resistance (HRS/OFF) RESET-state. For OxRAM devices, the conductive filament is composed of oxygen vacancies and defects [5], while in the case of CBRAM it consists of reduced metal ions sourced from a thin sacrificial metal anode [6]. For both CBRAM and OxRAM devices, the SET-state resistance (LRS) values can be well defined by controlling the dimensions of the conductive

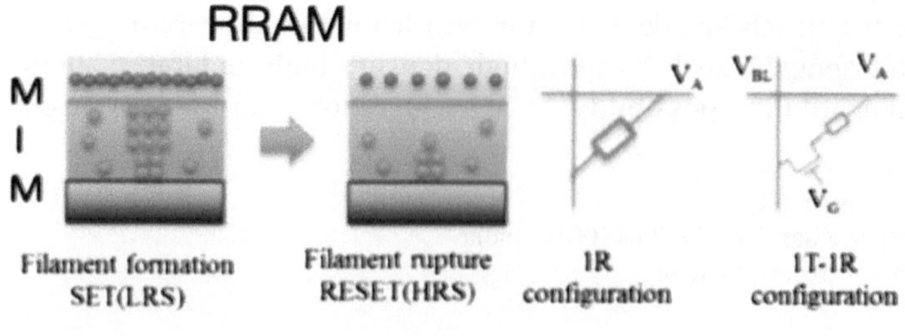

Fig. 1 Basic switching principle of filamentary RRAM devices

filament [2, 5], or the conduction mechanism (specific to some OxRAM devices) [7], which depends on the amount of current flowing through the active layer. Current flowing through the active layer is controlled either by externally imposed current compliance or by using an optional selector device that is used to drive the RRAM device (i.e., 1R-1T/1D configurations).

2.1 RRAM Resistance Variability

RESET-state resistance values (HRS) of filamentary RRAM devices generally present a large variable dispersion. The spread in HRS values arises due to stochastic breaking or uncontrolled dissolution of the conductive filament during the reset process. Different nanoscale factors, such as the presence of unavoidable defects close to the filament, interface effects, active-layer/electrode nonuniformity, material degradation/morphological changes on cycling, percolation paths, and process variations, may lead to HRS variability in cycle-to-cycle or device-to-device realizations [11–13]. Table 1 lists the HRS variability parameters (mean-μ, standard deviation-σ) for HfOx-based OxRAM, and Ag/GeS2-based CBRAM devices that were extracted from [2, 8], respectively. We chose different RESET programming conditions for HfOx-based OxRAM devices to study a wide range of mean HRS values. Device-to-device HRS spreads, for CBRAM and OxRAM, are shown in Fig. 2, generated by applying a log-normal distribution to the parameters listed in Table 1. While HRS variability profiles, like the ones shown in Fig. 2, are undesired for implementing multilevel memory states, in the following sections we show they can be exploited as an advantage in ELM architecture design (Fig. 3).

2.2 Cycle-to-Cycle Variability

Figure 4 and Table 3 present the evolution of cycle-to-cycle OFF/ON- state resistance variability with device active area dimensions. Large devices (active area of 1 ×

Table 1 Extracted HRS parameters for filamentary RRAM devices

Device, Technology, Reference	Mean (kΩ)	Variance (log10R)	Reset programming conditions (configuration)
CBRAM, (Ag/GeS2), [2]	892.86	0.6	VG = 2 V, VBL = 2 V, 10 s, (1R-1T)
OxRAM, (HfOx), [8]	25.12	0.03	−2.4 V, 50 ns (1R)
OxRAM, (HfOx), [8]	221.82	0.06	−2.7 V, 50 ns (1R)
OxRAM, (HfOx), [8]	2238.72	0.07	−3 V, 50 ns (1R)

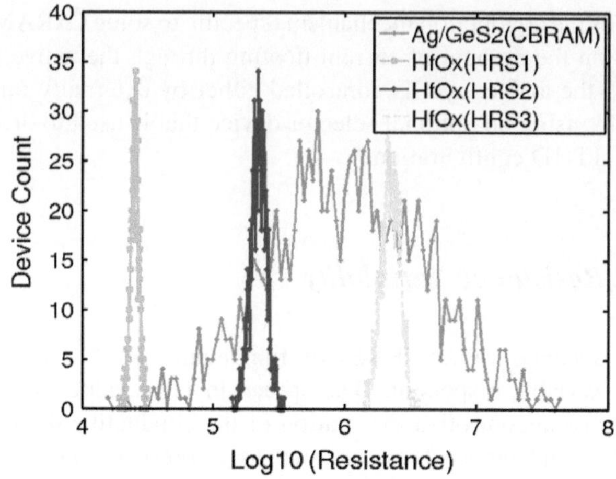

Fig. 2 Extracted HRS (R_{off}) log-normal distributions for filamentary RRAM devices listed in Table 1 [10]

Table 2 Extracted ON/OFF resistance spread parameters for TiO_2 and HfO_2 OxRAM devices [9]

Device, State	Mean (kΩ)	Standard Deviation ($log_{10}R$)
HfO_2 (On)	22.89	0.16
HfO_2 (Off)	3985.7	0.36
TiO_2 (On)	3.68	0.14
TiO_2 (Off)	193.5	0.26

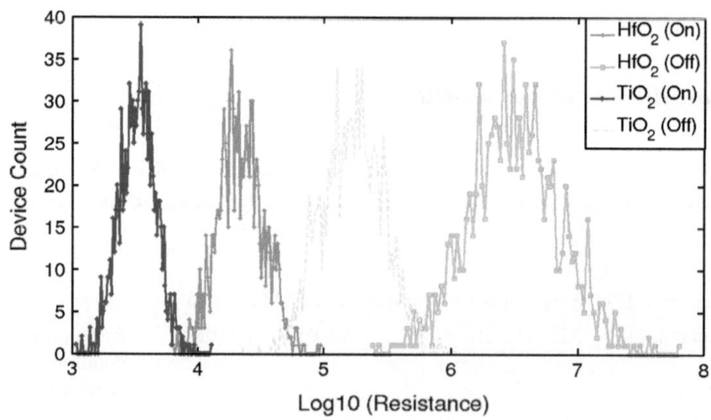

Fig. 3 Extracted (log-normal) ON and OFF resistance distributions for TiO_2 and HfO_2 OxRAM devices (parameters listed in Table 2) [9]

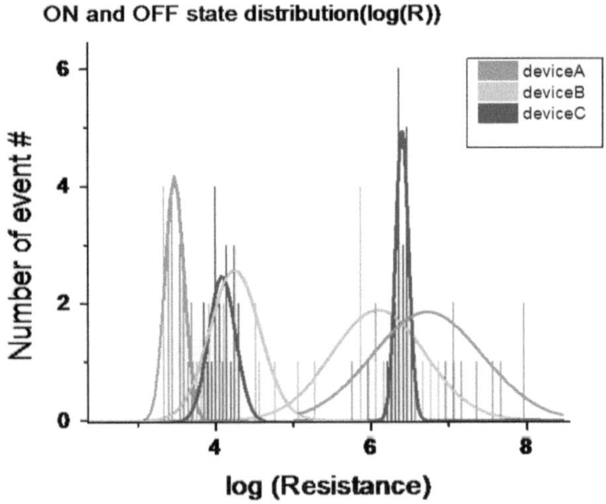

Fig. 4 ON/OFF-state resistance distribution for 20 cycles. VREAD = 0, 4 V. Devices A, B, C active areas are listed in Table 3. Spread increases with device active area dimensions

Table 3 Cycle-To-Cycle Resistance characteristics of the Fabricated Devices [14]

Device	Active Area	Thickness (nm)	State	Mean (log(R))	Std (log(R))
A	1μm × 1μm	10	On	3.47	0.11
A	1μm × 1μm	10	Off	6.72	0.69
B	500 nm × 500 nm	10	On	4.24	0.32
B	500 nm × 500 nm	10	Off	6.10	0.59
C	200 nm × 200 nm	10	On	4.08	0.17
C	200 nm × 200 nm	10	Off	6.40	0.08

1 μm^2) have a large variability, which decreases with the device dimensions. Such effect is notably due to the higher number of effective disrupted filaments (i.e., OFF paths), in the case of large devices that define the OFF-state. Unlike ON-state, which is mostly defined by the strongest filament, OFF-state is defined by multiple weak filaments obtained during forming and modified from cycle to cycle [14]. Stochastic filament formation and rupture leads to a spread in the IV traces and also RON/ROFF from cycle to cycle.

Figure 5 shows 30 cycle-to-cycle ON/OFF switching transitions and their corresponding reads at variable VREAD values (in the range of 0.1 to 1 V) for the same HfOx device. Between each RESET transition, the memory element is SET with identical positive voltage sweeps and current compliance. While ON-state spread almost stays constant with change in VREAD, the OFF-state variability significantly increases when VREAD is decreased. While such cycle-to-cycle ON/OFF-state resistance variability is a severe constraint for pure memory applications, we exploited this intrinsic effect for implementing stochasticity necessary for RBM architecture.

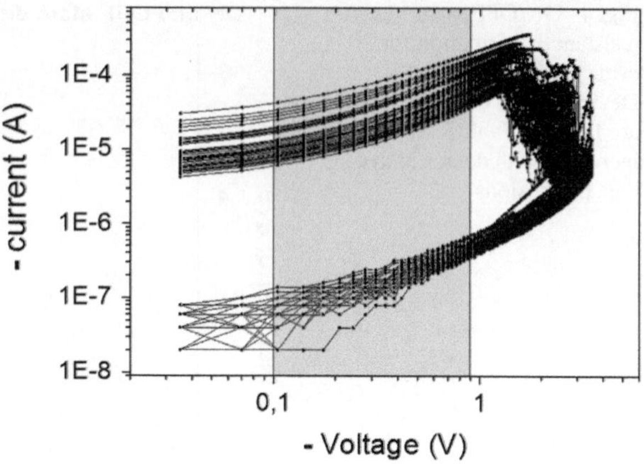

Fig. 5 30 successive RESET transitions for one OxRAM device. Change in ON/OFF-state dispersion with reference to change in VREAD is evidenced in the shaded region (0.1 to 1 V) [14]. (Note: VREAD does not modify the device state)

3 Extreme Learning Machine

3.1 Basics of Extreme Learning Machines (ELM)

An ELM basically consists of hidden layer synapses with randomly assigned weights, a hidden neuron layer with an infinitely differentiable activation function, and an output layer with synaptic weights determined by the learning rule shown in Fig. 6. Unlike other algorithms that try to assign hidden layer synaptic weights to some predetermined values while solving a QPP (Quadratic Programming Problem) for the output synaptic weights, (e.g., Support Vector Machines), or to improve them over successive iterations (e.g., backpropagation) [15], ELMs use random distributions of input weights and hidden layer neuron biases that remain fixed during learning [16]. Output synaptic weights in ELMs are determined through a simple L1-minimization scheme, i.e., using a matrix inversion [17]. Use of fixed random input layer weights allows the ELM to obtain a very good generalization behavior, compared to other gradient-based neural networks which explicitly try to tune all parameters [18]. ELMs simple learning algorithm gives it a strong advantage in terms of speed when compared to SVMs and other bioinspired algorithms [16].

3.2 Proposed OxRAM-ELM Architecture

Our proposed OxRAM-ELM architecture is shown in Fig. 7. The system is operated in training and test modes. Blockwise description is as follows:

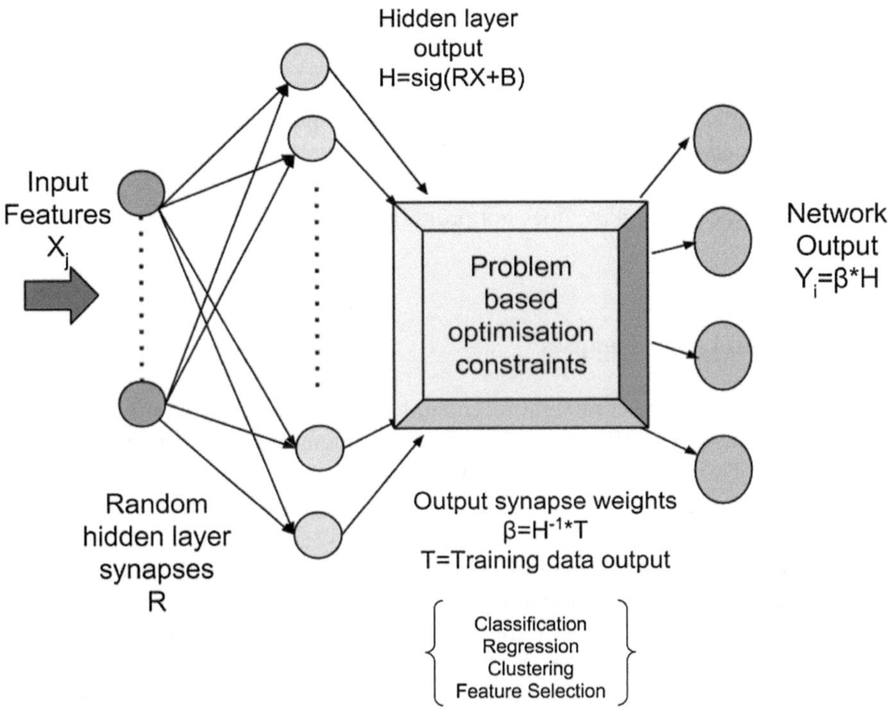

Fig. 6 Basic ELM framework and governing equations used to calculate output synaptic weights during training. [10]

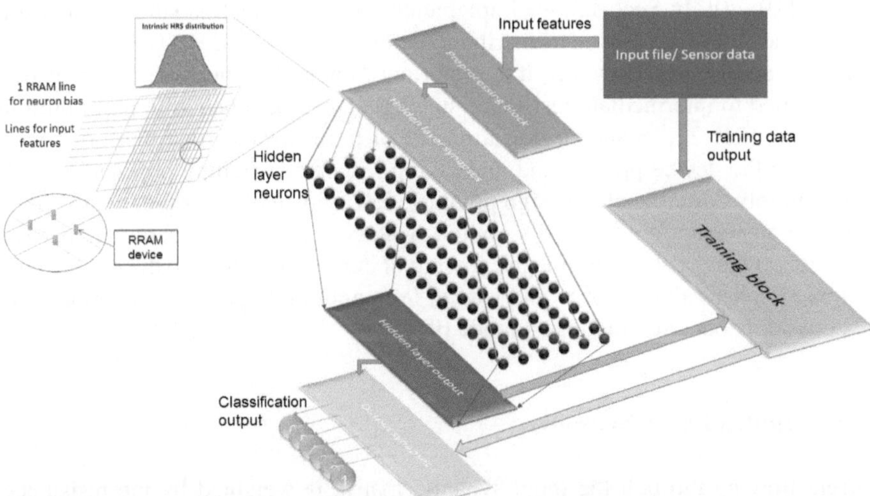

Fig. 7 Proposed OxRAM-ELM Architecture for multiclass classification. Resistance variability of OxRAM devices was exploited for implementing both random input weights and random neuron biases [9]

3.2.1 Preprocessing Block

Stored data are first preprocessed by conventional filtering, digital-to-analog conversion (DAC), and normalization steps. Output of the preprocessing block is in the form of voltage signals, fed directly to a network of hidden layer synapses with voltage amplitude below the switching threshold of the synaptic devices. This ensures that output of preprocessing stage does not program the input layer synapses, when not required [10].

3.2.2 Hidden Layer Synapses

In order to reach the higher integration capability, hidden layer synapses are realized using either a crossbar (1R) or matrix (1T-1R) configuration of filamentary (HfO_2) or interfacial (TiO_2) OxRAM devices. Such architecture is indeed perfectly adapted to the parallel structure of the hidden layer consisting N input neurons connected to M hidden neurons. Thus, the minimum size of the crossbar should be $(N + 1) \times M$; where N denotes the total number of input features (with one additional bias) and M denotes the total number of hidden layer neurons. ELM architectures require fixed random input weights. A RESET (or SET) operation is performed initially on the OxRAM synaptic matrix, before the launch of training mode, to obtain the type of OFF (or ON) state resistance distribution described in Sect. 2.1. The variable OFF (or ON) resistance spreads give rise to random input layer synaptic weights. Exploiting intrinsic resistance spreads is area and energy efficient as use of costly extrinsic techniques such as random number generator or PRNG (pseudo-random) circuits are avoided [19, 20]. In Sect. 3.3, we implemented the classification simulations using the extreme ON/OFF resistance distributions for both TiO_2 and HfO_2 devices (listed in Table 1 and Fig. 4). However, it is important to note that the devices can also be programmed to intermediate resistance states (i.e., can be tuned with programming conditions) [21, 22]. Standard deviation (σ) of the resistance distribution is more intrinsic to the device process and underlying resistive switching physics [13, 23]. From a purely circuit/analog design point of view, choice of μ for the input weights will have an impact on power dissipation, voltages/currents of operation, and line parasitic. μ would also determine the amount of current flowing into the hidden layer neurons. Higher values of μ would lead to less input layer power-dissipation when the network is operated in the test mode [9].

3.2.3 Hidden Layer Neurons

Current flowing through the input synaptic matrix is weighted by the resistances of the OxRAM devices and constantly integrated in the hidden layer neuron block. The ELM learning algorithm can work with many infinitely differentiable activation functions such as sine and radial basis function. We chose the sigmoid function (Fig. 8) as the hidden neuron for our architecture, based on the circuit implementation

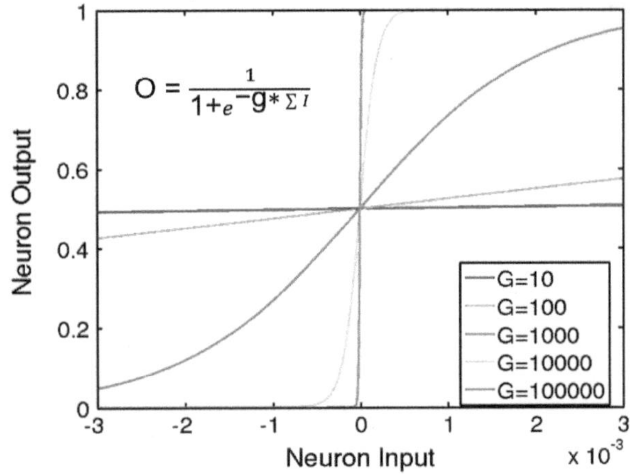

Fig. 8 Hidden layer neuron sigmoid response curve realized in the proposed OxRAM-ELM with different gain (g) values. Equation is also shown

described in [24]. The sigmoid response may be modified in the design by tuning the gain or slope, as shown in Fig. 8. Random neuron biasing can be achieved in an extremely area efficient manner by exploiting the hidden layer OxRAM synaptic matrix. (N + 1) th line in the OxRAM matrix (assuming the system has N input features) can be biased using a constant voltage source, applied across the top terminal of all the OxRAM devices. The bottom terminals of each individual device in the (N + 1)th line is directly connected to the individual hidden layer neurons. Since the resistance of all OxRAM devices in the matrix follows one of the resistance distributions shown in Fig. 2, the resultant current being fed into the hidden layer neurons from the (N + 1)th OxRAM line also follows a similar distribution. Such implementation avoids the need of any external bias randomization circuit.

3.2.4 Training Block

This block is active only during the training-mode operation of the network. In the training mode, data are simultaneously fed to the preprocessing and the training blocks. For each training data point, the output of the hidden layer neurons and the expected output are stored inside the training block (see Fig. 9). Thus, the minimum size of training block memory is given by the following expression. Once all hidden layer neuron responses and their corresponding output values/classes have been stored in the training module memory, a matrix inversion is performed as described in [16], in order to solve the linear system of equations and generate the synaptic weights for the output layer. The matrix inversion can be implemented using algorithms such as Gauss–Jordan elimination or QR decomposition using the tool discussed in [25] or through LU factorization technique [26].

$$\begin{aligned}\textit{Training block memory} &= \#\textit{Training samples} \times (\#\textit{Hidden neurons} + 1) \\ &\times \textit{Data bit width}\end{aligned} \tag{1}$$

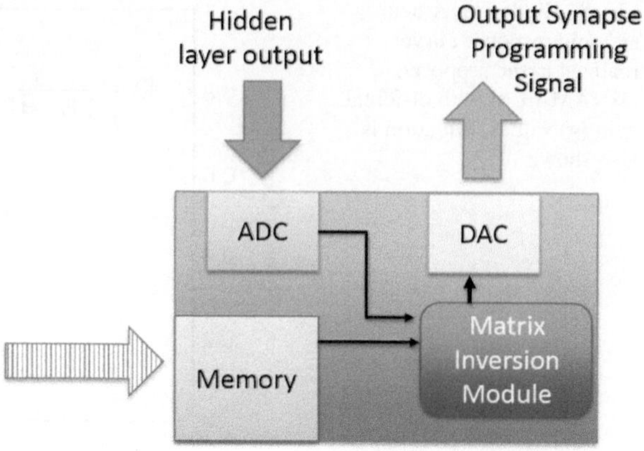

Fig. 9 Training block schematic

3.2.5 Output Layer Synapses

These are programmed only at the end of the training mode. As the output layer weights are generated from a matrix inversion operation, they will have a wide dynamic range. We choose to implement the output weights as ideal synapses (i.e., using purely digital memory). Floating point representation enables to store values of a very high dynamic range by using fewer bits. Hence, we used a 12-bit floating representation (2 bits mantissa and 10 bits exponent). This also helped with the problem of realizing +ve and −ve output synaptic weights.

3.2.6 Output Neuron

In contrast to the hidden layer, which uses a sigmoid activation function, for the output neuron we make use of a linear activation function. This is primarily due to the reason that we make use of a linear equation solver to find/calculate the output synaptic weights. For the multiclass classification, we require multiple output neurons (one per class). During the training phase, the output layer is switched off in order to minimize power dissipation.

3.3 Results

3.3.1 Image Segment Classification

The goal of this test was to classify segments of images into 7 classes (listed in Table 3) based on the image pixel corresponding to it and its neighborhood. Features used for classification are described in [27] simulated OxRAM-ELM network consists of 19 input nodes (1 per input features), 4000 hidden synapses, 200 hidden neurons, 1400

Table 4 Image segment labels

Sr.no.	Class of image segment
1	Brickface
2	Sky
3	Foliage
4	Cement
5	Window
6	Path
7	Grass

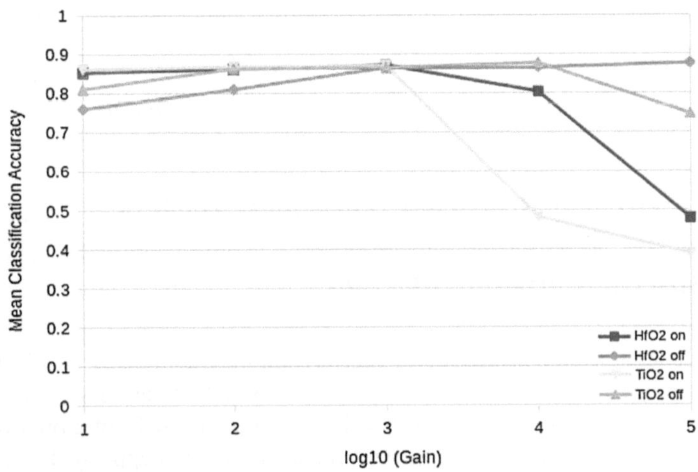

Fig. 10 Dependency of mean classification accuracy on device resistance distributions and hidden layer neuron gain (mean obtained on 10 cycles for each point)

output stage synapses, and 7 output neurons (1 per class). Figure 10 shows the mean image segment classification accuracy over 10 test cycles for all device resistance distributions listed in Table 2 with varying hidden layer neuron gain values. The comparison is also listed in Table 4.

3.3.2 Sinc Function Regression

The goal of this experiment was to emulate the functionality of a Sinc function generator using the RRAM-ELM architecture. The data provided to the network were comprised of uniformly distributed samples of input and output values of an ideal Sinc function over the interval of −10 to 10. Training data size as well as test data size was 5000 sample points. The simulated network consists of 1 input node, 40 hidden layer synapses, 20 hidden layer neurons, 20 output layer synapses, and 1 output neuron (Fig. 11).

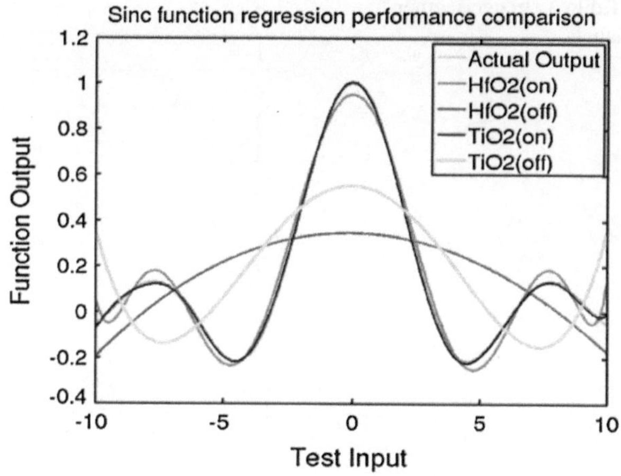

Fig. 11 Sinc Regression Performance Comparison for RRAM-ELM

4 CMOS-RRAM Restricted Boltzmann Machine

4.1 Restricted Boltzmann Machines

RBM is a generative, stochastic neural network architecture consisting of two layers of nodes representing visible and hidden variables. RBMs are a variant of Boltzmann machines, with the restriction that their neurons must form a bipartite graph [28]. There are weighted connections between every node in opposite layers, and no connections between any nodes in the same layer (Fig. 12). The following notation system will be used: vi and hj are the binary states of the ith and jth node, where i = 1 .. I and j = 1,..,J, in the visible and hidden layer, respectively; w_{ij} is the connection weight between the ith and jth nodes.

4.1.1 Mathematical Formulation

The energy function E (v, h) of an RBM is defined as:

$$Energy(v, h) = -b'h - c'v - h'Wv. \tag{2}$$

Here W represents the weights connecting hidden and visible units, and b, c are the offsets of the visible and hidden layers, respectively. This translates directly to the following free energy formula adapted from [28]:

$$FreeEnergy(v) = -b'v - \sum_i \log \sum_{h_i} e^{h_i(c_i+W_i v)} \tag{3}$$

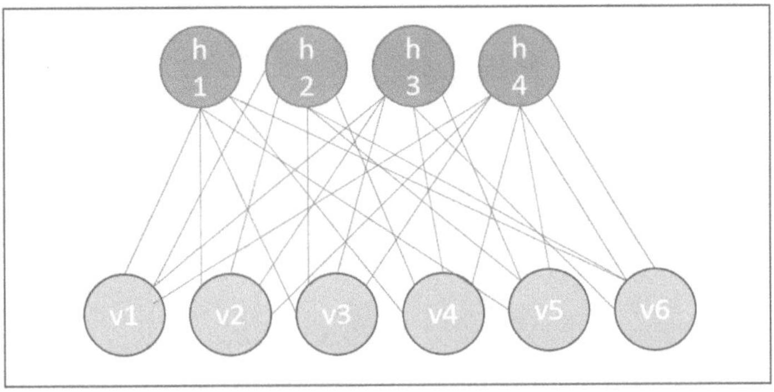

Fig. 12 General RBM architecture. h denotes hidden layer nodes, while v denotes visible layer nodes. There are no connections within the same layer

Because of no lateral connections within a given layer (definition of RBM), neurons are conditionally dependent only on the other layer, and independent of entities within the same layer. Using this property, we can write:

$$p(v|h) = \prod_i p(h_i|v)) \tag{4}$$

$$p(h|v) = \prod_i p(v_i|h)) \tag{5}$$

4.1.2 Training Algorithm

Alternating Gibbs Sampling (AGS) and Contrastive-Divergence learning (CD) have been found to be an effective process to determine the node states and update the weight parameters [29], respectively. For the proposed RRAM-based implementation, we make use of contrastive divergence. The steps followed in the algorithm are listed as:

- Take a training sample v, compute the probabilities of the hidden units and sample a hidden activation vector h from this probability distribution.
- Compute the vector product of v and h; termed as the positive gradient.
- From h, sample a reconstruction v' of the visible units, then resample the hidden activations h' from this. (Gibbs sampling step)
- Compute the outer product of v' and h'; termed as the negative gradient.
- The weight update to w_{ij} is the positive gradient minus the negative gradient, times learning rate (accuracy optimization parameter) [28, 29]:

$$\Delta\, w_{ij} = \epsilon \times (v \cdot h - v' \cdot h') \tag{6}$$

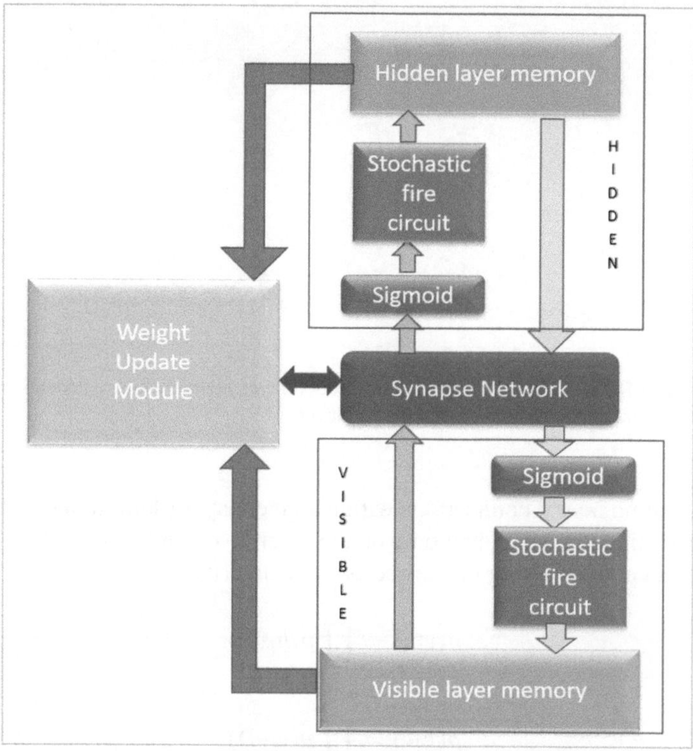

Fig. 13 Proposed hybrid RRAM-CMOS RBM architecture with individual functional sub-blocks. RRAM devices are used in **a** Synapse, **b** Hidden/Visible memory, and **c** Stochastic fire circuit blocks [14]

4.2 Proposed RBM Architecture

Our proposed RBM architecture is shown in Fig. 13 [14]. The system consists of 2 layers of stochastic sigmoid neurons (hidden and visible), all fully connected. For classification, a third layer is also added. Individual functional sub-blocks are realized as follows:

4.2.1 Synapse Network

The synaptic array can be realized with either a crossbar or a matrix of RRAM devices. In the proposed scheme, each synapse is realized using 4 binary RRAM devices (to obtain a 4-bit weight resolution). This scheme allows us to have a reasonable amount of weight accuracy while keeping the hardware complexity and area requirements manageable. The 4-device/synapse approach can be further simplified, if the RRAM device being used offers multilevel programming capability. Our fabricated HfOx devices offered good binary programming with a conservative resistance window.

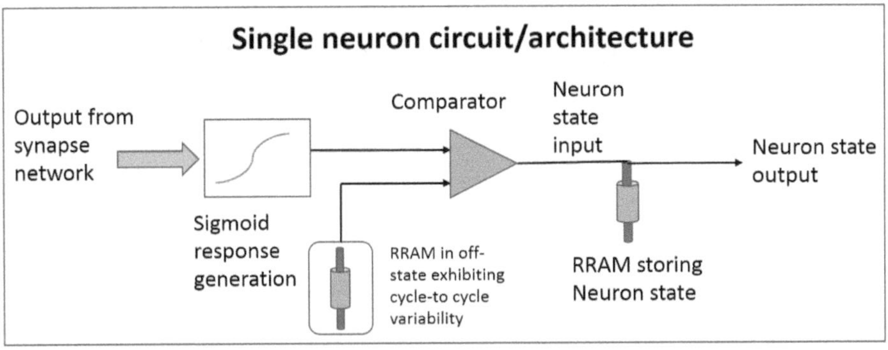

Fig. 14 Block-level design of a single neuron. Consists of sigmoid circuit, RRAM devices for stochastic activation and neuron internal state storage, and comparator

4.2.2 Stochastic Neuron

Figure 14 shows the full stochastic neuron block. Each neuron (hidden or visible) follows a sigmoid response, which is implemented by a low-power 6-T CMOS sigmoid circuit. The gain of the sigmoid circuit can be tuned by optimizing the scaling of the six transistors [30]. Voltage output of sigmoid circuit is compared with the voltage drop across an RRAM device, with the help of a comparator. The HfOx-based device is repeatedly cycled to OFF-state. The cycle-to-cycle intrinsic R_{ON} and R_{OFF} variability of the RRAM device leads to a variable reference voltage for the comparator. This helps translate the deterministic sigmoid output to an overall neuron output, which is stochastic in nature. At any given moment, the specific neuron output is also that specific neurons internal state, which needs to be stored for RBM-driven learning. The internal state of each neuron is stored using individual RRAM devices, placed after the comparator. 1-RRAM device/neuron is sufficient since RBM requires each neuron to have either 1 or 0 as its binary activation state.

4.2.3 CD Weight Update Block

The weight update module is a purely digital circuit that accesses synaptic weights and internal neuron states. It updates the synaptic weights during learning by applying the contrastive-divergence RBM algorithm. The block consists of an array of weight update circuits, one of which is shown in Fig. 15. Synaptic weight is updated (by $\triangle$Wij), based on the previous (v, h) and current (v, h) internal neuron states of the mutually connected neurons of the hidden and visible layers. CD is realized using two AND gates and a tri-state comparator (having outputs $-1, 0, +1$). Input to the first AND gate is previous internal neuron states, while the input to second AND gate is the current internal neuron states. Based on the tri-state comparator output (learning rate- optimization parameter) will either be added, or subtracted, or no-change will be made to the existing synaptic weight (w_{ij}).

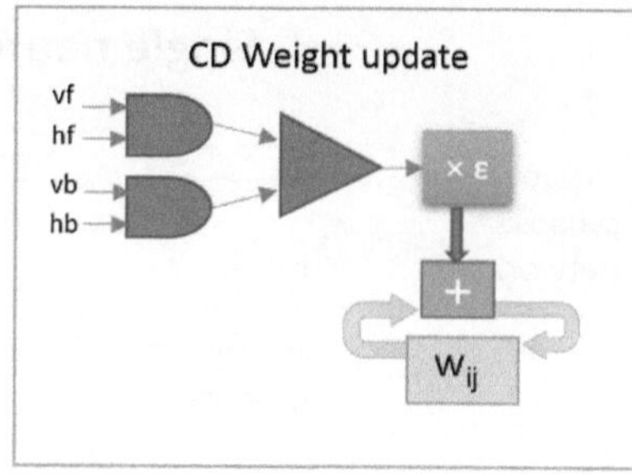

Fig. 15 Training block

4.3 Results and Discussion

4.3.1 Learning Performance

Simulation results are shown in Fig. 16 and Table 5. Average learning accuracy of the system, for both R_{ON} and R_{OFF} distributions, was 89 %. From Tables 3 and 5, it is evident that exploitation of R_{ON} distributions in the stochastic neuron activation block gave a higher learning accuracy compared to R_{OFF}, for all device dimensions. This may possibly happen because resistance variation in ON-state is in the order of kΩ, compared to MΩ in the case of OFF-state. Since the cycled OxRAM resistances are eventually mapped and compared against a small voltage range (i.e., the V_{neuron}), 0 to 1 V, the OxRAM reference voltage points are more evenly distributed in the case of ON-state and tend to get saturated for OFF-state resistance distributions. Note from Table 5 that ϵ was used as an optimization parameter for improving the learning accuracy. Figure 16b shows that reconstruction error for the reduced MNIST dataset minimizes after 80 epochs (Tables 6 and 7).

4.3.2 Stochastic Neuron Activation Block

The neuron circuit was simulated in cadence using 90-nm CMOS design kit (Fig. 17). Conductive-filament (CF)-based Verilog-A compact model proposed in [31] was used to model our HfOx devices. The model parameters were tweaked to emulate RON/ROFF spreads equivalent to the ones shown in Fig. 4 and Table 3. The circuit shown in Fig. 17 has two subcircuits: (a) 6-T sigmoidal function circuit and (b) OxRAM-comparator circuit. Pre-synaptic input to the sigmoid circuit is modeled by a current source (I_{dc}), while the output is a voltage (V_{neuron}), thus making it a resistive type of neuron [30]. Value of V_{neuron} depends whether the pre-synaptic input acts as a current sink (−ve I_{dc}) or current source (+ve I_{dc}). The second subcircuit consists of an OxRAM device connected to two voltage sources, V_{pulse1} and V_{pulse2} (used to generate set, reset and read bias voltages). A read voltage V_{READ} (V_{pulse1} V_{pulse2}) and series resistance R are used to obtain a voltage drop (V_{oxram}) for the reference-voltage input of the comparator. Due to the low value of read current,

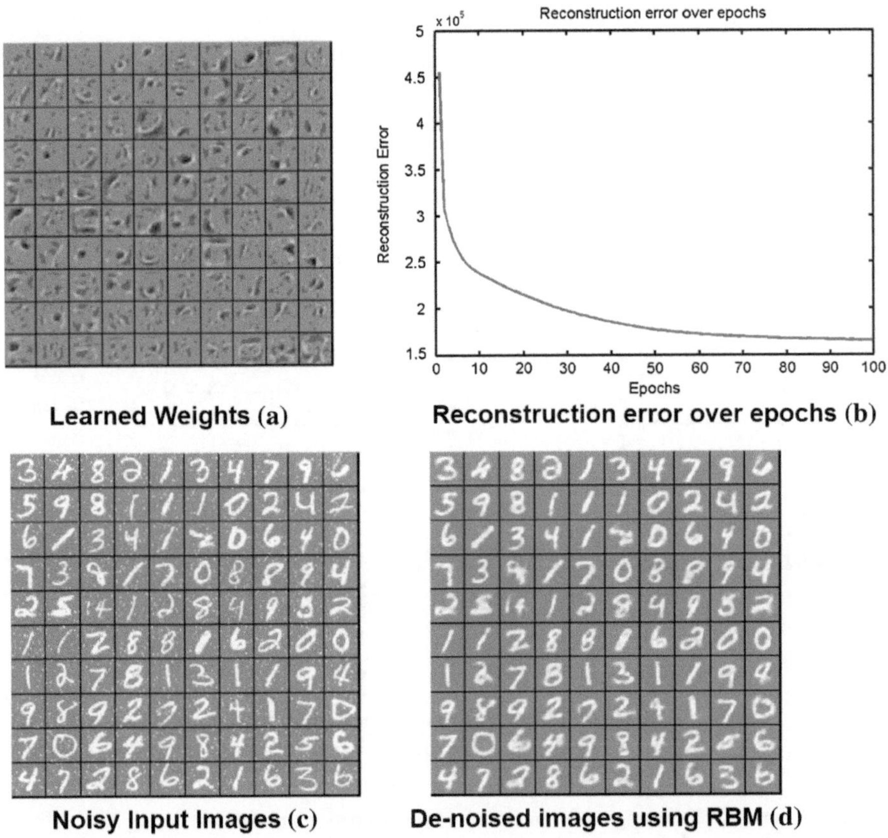

Fig. 16 Simulation results of proposed RBM on reduced MNIST dataset of 6000 images

Table 5 Image segmentation performance for the proposed ELM architecture

Synapse type	Test acc (mean)	Test acc (std dev)
Ideal ELM	94.84	0.81
HfO2(On)	95.65	0.26
HfO2(Off)	95.31	0.48
TiO2(On)	95.07	0.37
TiO2(Off)	95.58	0.34

Table 6 Sinc regression performance comparison for RRAM-ELM

Device (State)	Mean square error
HfO_2 (On)	0.0409
HfO_2 (Off)	0.3154
TiO_2 (On)	0.0071
TiO_2 (Off)	0.2576

Table 7 Performance results of the system

	State	Test accuracy	Epochs	ϵ
Device A	On	91.9	100	0.08
	Off	88.6	100	0.04
Device B	On	89.1	25	0.08
	Off	88.9	100	0.08
Device C	On	92.2	100	0.08
	Off	83.8	100	0.04

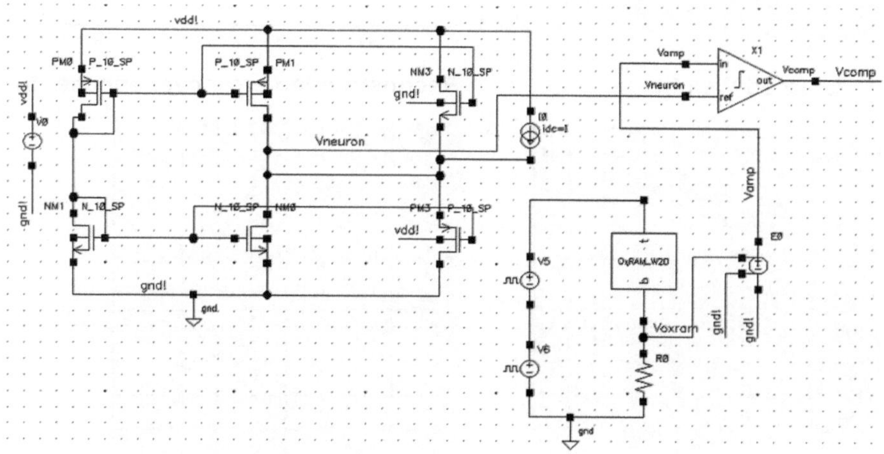

Fig. 17 Schematic of the single neuron circuit used in Cadence simulations

V_{oxram} needs to be amplified before it is fed to the comparator. Based on the V_{neuron} and amplifier output, the comparator generates logic values 1 s or 0 s for each run of the simulation. Cycle-to-cycle variability of R_{OFF} also makes the comparator reference-voltage variable, thus leading to a stochastic comparator output. Figure 18 shows the distribution of the stochastic neuron block output as a function of the comparator inputs (V_{neuron} & V_{oxram}), simulated for 100 points. VREAD = 0.1V, $R = 5\ k\Omega$, and an amplifier gain of 350 was used for exploiting the ROFF distribution of device B. However, if R_{ON} distribution is used, value of R and gain should be changed accordingly. From Fig. 18, it is evident that majority of the activations happen when the OxRAM resistance attains lower values of R_{OFF}.

4.3.3 Endurance Estimation

Average switching activity for HfOx devices (inclusive of all 3 applications—synaptic, stochastic neuron activation, and internal neuron state storage) was found to be around ∼14 million events per epoch from full system simulations. The max-

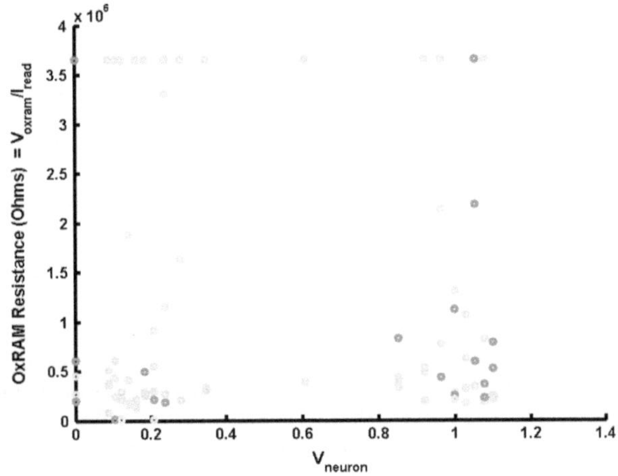

Fig. 18 Distribution of stochastic neuron block output as a function of comparator inputs (V_{neuron} and V_{oxram}). Majority of activations occur when the cycled OxRAM device attains low-resistance values

imum switching activity for any single OxRAM device comes out ~2 million for the full training simulation. This number is further expected to go up as the number of data points in the training set increases. Thus for large datasets, memory devices with strong endurance characteristics are required. HfOx-based devices can satisfy this constraint easily [14].

5 Conclusion

As shown in the previous two sections, we have shown how two different methods utilize variability for two completely different types of learning algorithms.

ELM is a simple neural network with supervised learning utilizing random hidden nodes in order to achieve universal approximation. Hence, it required static variability which we utilized in the form of device-to-device variability of RRAM. Thus, we have utilized noise/faults to achieve computation.

On the other hand, RBM is an unsupervised learning algorithm using stochastic neuron activation to model neural activity. This can be modeled using dynamic (cycle to cycle) variability of RRAM. Also it forms the unit cell of deep neural network. In this case, we have utilized the unreliability at low current to achieve computation.Also notable is the fact that here RRAM was primarily utilized as a digital device in contrast to its use in the ELM design where its use was completely as an analog device.

In both cases, modeling the respective variability using standard digital components would be possible but would cause high power dissipation and area usage.

References

1. Suri, M., Bichler, O., Querlioz, D., Cueto, O., Perniola, L., Sousa, V., Vuillaume, D., Gamrat, C., DeSalvo, B.: Phase change memory as synapse for ultra-dense neuromorphic systems: Application to complex visual pattern extraction. 2011 IEEE International Electron Devices Meeting (IEDM), vol. 4, no. 4, pp. 5–7, Dec 2011
2. Suri, M., Querlioz, D., Bichler, O., Palma, G., Vianello, E., Vuillaume, D., Gamrat, C., DeSalvo, B.: Bio-inspired stochastic computing using binary cbram synapses. IEEE Trans. Electron Devices **60**(7), 2402–2409 (2013)
3. Vincent, A.F., Larroque, J., Zhao, W.S., Romdhane, N.B. Bichler, O., Gamrat, C., Klein, J.O., Galdin-Retailleau, S., Querlioz, D.: Spin-transfer torque magnetic memory as a stochastic memristive synapse. In: Circuits and Systems (ISCAS), vol. 2014, pp. 1074–1077 (2014)
4. DeSalvo, B., Sousa, V., Perniola, L., Jahan, C., Maitrejean, S., Nodin, J., Cagli, C., Jousseaume, V., Molas, G., Vianello, E. et al.: Emerging memory technologies: Challenges and opportunities
5. Wong, H.-S.P., Lee, H.-Y., Yu, S., Chen, Y.-S., Wu, Y., Chen, P.S. Lee, B., Chen, F.T., Tsai, M.-J.: Metal–Oxide RRAM, vol. 100, no. 6. IEEE, pp. 1951–1970 (2012)
6. Gopalan, C., Ma, Y., Gallo, T., Wang, J., Runnion, E., Saenz, J., Koushan, F., Blanchard, P., Hollmer, S.: Demonstration of conductive bridging random access memory (cbram) in logic cmos process. Solid-State Electron. **58**, 1 (2011)
7. Su, Y.-T., Chang, K.-C., Chang, T.-C., Tsai, T.-M., Zhang, R., Lou, J., Chen, J.-H., Young, T.-F., Chen, K.-H., Tseng, B.-H., et al.: Characteristics of hafnium oxide resistance random access memory with different setting compliance current. Appl. Phys. Lett. **103**(16), 163502 (2013)
8. Yu, S., Guan, X., Wong, H.-S.P.: On the switching parameter variation of metal oxide rrampart ii: model corroboration and device design strategy. IEEE Trans. Electron Devices **59**(4), 1183–1188 (2012)
9. Suri, M., Parmar, V., Sassine, G., Alibart, F.: Oxram based elm architecture for multi-class classification applications. Neural Netw. (IJCNN) **2015**, 1–8 (2015)
10. Suri, M., Parmar, V.: Exploiting intrinsic variability of filamentary resistive memory for extreme learning machine architectures. IEEE Trans. Nanotechnol. **14**(6), 963–968 (2015)
11. Raghavan, N.: Performance and Reliability Trade-offs for High-κ RRAM. Elsevier, vol. 54, no. 9 (2014)
12. Fantini, A., Goux, L., Degraeve, R., Wouters, D., Raghavan, N., Kar, G., Belmonte, A., Chen, Y.-Y., Govoreanu, B., Jurczak, M.: Intrinsic Switching Variability in Hfo 2 RRam, pp. 30–33 (2013)
13. Chen, A., Lin, M.-R.: Variability of resistive switching memories and its impact on crossbar array performance. Reliab. Phys. Symp. (IRPS) **2011** (2011)
14. Suri, M., Parmar, V., Kumar, A., Querlioz, D., Alibart, F.: Neuromorphic hybrid rram-cmos rbm architecture. In: 2015 15th Non-Volatile Memory Technology Symposium (NVMTS), pp. 1–6, Oct 2015
15. Huang, G.-B.: An insight into extreme learning machines: random neurons, random features and kernels. Cogn. Comput. 1–15 (2014)
16. Huang, G.-B., Zhu, Q.-Y., Siew, C.-K.: Extreme learning machine: theory and applications. Neurocomputing **70**(1), 489–501 (2006)
17. Yang, A.Y., Sastry, S.S., Ganesh, A., Ma, Y.: Fast l 1-minimization algorithms and an application in robust face recognition: A review, pp. 1849–1852 (2010)
18. Huang, G.-B., Zhu, Q.-Y., Siew, C.-K.: Extreme learning machine: a new learning scheme of feedforward neural networks. Neural Netw. **2004**, 985–990 (2004)
19. Merkel, C., Kudithipudi, D.: A current-mode cmos/memristor hybrid implementation of an extreme learning machine. In: Proceedings of the 24th edition of the Great Lakes Symposium on VLSI, pp. 241–242. ACM (2014)
20. Decherchi, S., Gastaldo, P., Leoncini, A., Zunino, R.: Efficient digital implementation of extreme learning machines for classification. IEEE Trans. Circuits Syst. II: Express Briefs **59**(8), 496–500 (2012)

21. Lee, H.Y., Chen, P.S., Wu, T.Y., et al.: Low power and speed bipolar switching with a thin reactive buffer layer in robust HfO2 based RRAM. Int. Electron Devices Meet. (2008)
22. Su, Y.-T., et al.: Characteristics of hafnium oxide resistance random access memory with different setting compliance current. Appl. Phys. Lett. **103**, 16 (2013)
23. Yu, S., Guan, X., Wong, H.-S.P.: On the stochastic nature of resistive switching in metal oxide rram: physical modeling, monte carlo simulation, and experimental characterization. Int. Electron Device Meet. (2011)
24. Shi, B., Chen, L., Lu, C.: Current controlled sigmoid neural circuit. U.S. Patent, vol. 6, Dec 2003
25. Irturk, A., Benson, B., Mirzaei, S., Kastner, R.: An fpga design space exploration tool for matrix inversion architectures. Appl. Specif. Processors **2008**, 42–47 (2008)
26. Zhang, W., Betz, V., Rose, J.: Portable and scalable fpga-based acceleration of a direct linear system solver. In: ACM Transactions on Reconfigurable Technology and Systems (TRETS), vol. 5, p. 1 (2012)
27. Bache, K., Lichman, M.: UCI Machine Learning Repository [Irvine. CA: University of California, School of Information and Computer Science (2013). http://archive.ics.uci.edu/ml
28. Hinton, G.: A practical guide to training restricted boltzmann machines. Momentum **9**(1), 926 (2010)
29. Hinton, G.E., Osindero, S., Teh, Y.W.: A fast learning algorithm for deep belief nets. Neural Comput. **18**(7), 1527–1554 (2006)
30. Pan, D., Wilamowski, B.M.: A vlsi implementation of mixed-signal mode bipolar neuron circuitry. In: Proceedings of the International Joint Conference on Neural Networks, 2003, vol. 2, pp. 971–976, July 2003
31. Li, H., Jiang, Z., Huang, P., Wu, Y., Chen, H.Y., Gao, B., Liu, X.Y., Kang, J.F., Wong, H.S.P.: Variation-aware, reliability-emphasized design and optimization of rram using spice model. In: 2015 Design, Automation Test in Europe Conference Exhibition (DATE), pp. 1425–1430, Mar 2015

Theoretical Analysis of Spike-Timing-Dependent Plasticity Learning with Memristive Devices

Damien Querlioz, Olivier Bichler, Adrien F. Vincent and Christian Gamrat

Abstract Several recent works, described in chapters of the present series, have shown that memristive devices can naturally emulate variations of the biological spike-timing-dependent plasticity (STDP) learning rule and can allow the design of learning systems. Such systems can be built with memristive devices of extremely diverse physics and behaviors and are particularly robust to device variations and imperfections. The present work investigates the theoretical roots of their STDP learning. It is suggested, by revisiting works developed in the field of computational neuroscience, that STDP learning can approximate the machine learning algorithm of Expectation-Maximization, the neural network operation implementing "Expectation" steps, while STDP itself implementing "Maximization" steps. This process allows a system to perform Bayesian inference among the values of a latent variable present in the input. This theoretical analysis allows interpreting how STDP differs for several device physics and why it is robust to device mismatch. It can also provide guidelines for designing STDP-based learning systems.

1 Introduction

In recent years, memristive devices have emerged as an attractive opportunity to implement synapses in neuroinspired systems [6, 7, 9–11, 14, 20, 23, 26, 28, 29]. Memristive devices can indeed provide a form of compact embedded and nonvolatile

D. Querlioz (✉) · A.F. Vincent
CNRS, Institut d'Electronique Fondamentale, University of Paris-Sud, Orsay, France
e-mail: damien.querlioz@u-psud.fr

A.F. Vincent
e-mail: adrien.vincent@u-psud.fr

O. Bichler · C. Gamrat
CEA, LIST, Saclay, France
e-mail: olivier.bichler@cea.fr

C. Gamrat
e-mail: christian.gamrat@cea.fr

© Springer (India) Pvt. Ltd. 2017

M. Suri (ed.), *Advances in Neuromorphic Hardware Exploiting Emerging Nanoscale Devices*, Cognitive Systems Monographs 31,
DOI 10.1007/978-81-322-3703-7_10

memory, which can be used as a binary memory and sometimes even as a multilevel or partly analog memory.

Additionally, a promising lead is to not use the memristive devices solely as memory, but to harness their physics to implement learning rules, in similarity to real biological synapses. In this context, the most investigated learning approach is based on spike-timing-dependent plasticity (STDP). This learning rule, which is inspired by biological measurements performed since the late 1990s [2, 16], can be implemented by memristive nanodevices directly [1, 12, 13, 15, 25, 27, 33] or under simplified forms [3, 21–23, 31, 32]. The capabilities of STDP have mostly been investigated within neuroscience studies [17, 18]. It has also been shown that a system equipped with STDP can learn advanced proof-of-concept tasks [3, 21]. Nevertheless, if we are to develop useful systems exploiting nanodevice-based STDP, a deeper understanding of the computational power of STDP-based learning is necessary. Additionally, it is essential to understand how nanodevice-specific questions affect the STDP-learning process.

The present chapter starts to address these questions. It builds on theoretical works originally introduced in the context of neuroscience in Ref. [19]. These works interpret STDP-trained systems as performing a form of Bayesian inference. They also show that STDP learning approximates the algorithm of Expectation-Maximization, a powerful machine learning algorithm. The present chapter adapts these theoretical works to nanodevices in the context of the simplified STDP approach of [3, 21, 22, 31, 32]: In particular, the chapter highlights how the different device physics (presented in the other chapters of this book) affect STDP learning and its interpretation as Bayesian inference. This theoretical analysis also allows us to discuss and interpret specific questions related to nanodevices, such as the impact of device variations.

The present chapter extends a discussion originally appearing in Ref. [23].

2 Spike-Timing-Dependent Plasticity and Expectation-Maximization

2.1 *Simplified STDP*

Most STDP models are inspired by biological measurements or considerations. However, in an important work, Nessler et al. showed that STDP can lead to a form of optimal Bayesian inference [19], by approximating the powerful machine learning algorithm of Expectation-Maximization. For this demonstration, the authors employ a highly simplified version of STDP, which deviates from most models of STDP in several aspects. In conventional STDP, conductance change occurs when a pair of spikes occur at near times on the two sides (presynaptic and postsynaptic) of a synapse. This is illustrated in the conventional STDP graph of Fig. 1a.

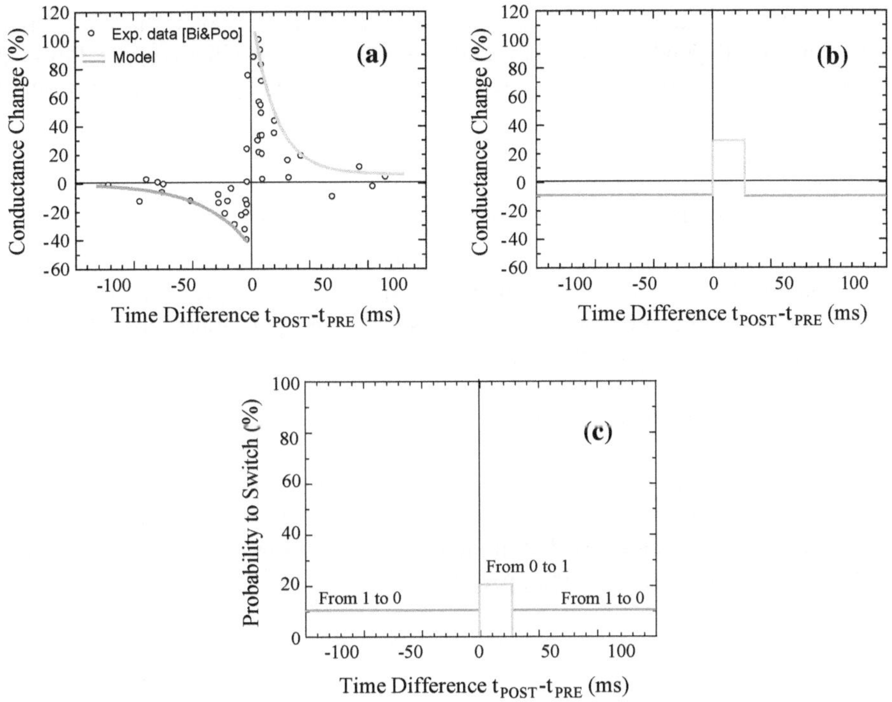

Fig. 1 Illustration of several STDP rules. **a** Original measurement from biology (symbols) [2] and the conventional model of STDP (full lines). **b** Simplified STDP used within the present chapter. **c** Stochastic version of the simplified STDP

By contrast, in the approach of Ref. [19], conductance changes occur specifically when a postsynaptic spike occurs. After the postsynaptic neuron of a synapse has spiked:

1. if there was a presynaptic spike recently, within a STDP "window," the weight of the synapse increases by a weight increment δw_+
2. if there was no recent presynaptic spike, the weight of the synapse decreases by a weight decrement δw_-.

This simplified rule maps especially well to nanodevices [3, 21, 23, 31, 32], and we use variations of it throughout the whole chapter. It is possible to plot this STDP rule on a graph similar to conventional STDP (Fig. 1b), although this plot might be misleading in that STDP does not specifically occur when a pair of spikes occur.

As equations for the weight increment δw_+ and the weight decrement δw_-, Nessler et al. considered:

$$\delta w_+ = C \exp(-w) - 1$$
$$\delta w_- = 1, \tag{1}$$

where C is a real constant greater than one. It should be noted that the weight increment δw_+ depends highly on the current conductance w of the synapse. Physical memristive devices, of course, do not obey these particular equations. The impact of their device physics will be discussed in Sect. 3.

2.2 Connection with Expectation-Maximization

Based on Eq. (1), it is possible to analyze formally the learning process. In particular, we can derive analytically the value that the synaptic weights approach at the end of a learning process. We can introduce

$$p(PRE|POST) = p(t_{PRE} \in [t_{POST} - t_{STDP}; t_{POST}]|t_{POST}), \tag{2}$$

which represents, for a given synapse, the probability that when its postsynaptic neuron spiked, its presynaptic neuron spiked in the STDP window t_{STDP} preceding the spike. Then, we can show that the final weight w_∞ that this particular synapse will approach during the learning process is

$$w_\infty = \log p(PRE|POST) + \log C. \tag{3}$$

(log represents the natural logarithm). The derivation of this equation appears in the Appendix of the chapter.

This result has deep implications. Nessler et al. have shown that this allows a system with simplified STDP to perform an approximation of Expectation-Maximization, a powerful machine learning algorithm [8]. The process is illustrated in Fig. 2. We consider a feedforward system, where input neurons are connected to output leaky-integrate-and-fire neurons, in an all-to-all manner, by synapses equipped with simplified STDP. The crossbar shown in Fig. 2 naturally implements this system. Additionally, when an output neurons spike, it inhibits all the other output neurons. The system is therefore reminiscent of a winner-takes-all architecture. The system has two distinct behaviors. Most of the time, the output neurons integrate information, until an output neuron spikes. This corresponds to an Expectation step of the Expectation-Maximization algorithm. After the output neuron has spiked, the weights of the synapses are updated according to the simplified STDP rule. This corresponds to a Minimization step of the Expectation-Maximization algorithm.

This correspondence can explain that this system has been used successfully for learning different tasks like handwritten character digit recognition [21], car detection in a video [4], or audio pattern recognition [31]. In each case, the system, using the Expectation-Maximization analog, is capable of identifying the latent variable behind the inputs and of classifying input along the different identified latent variable values.

These theoretical considerations provide insight into how our inference engine learns and performs inference. However, physical memristive synapses do not lead to a behavior equivalent to Eq. (1). We now consider the details regarding the importance of this difference.

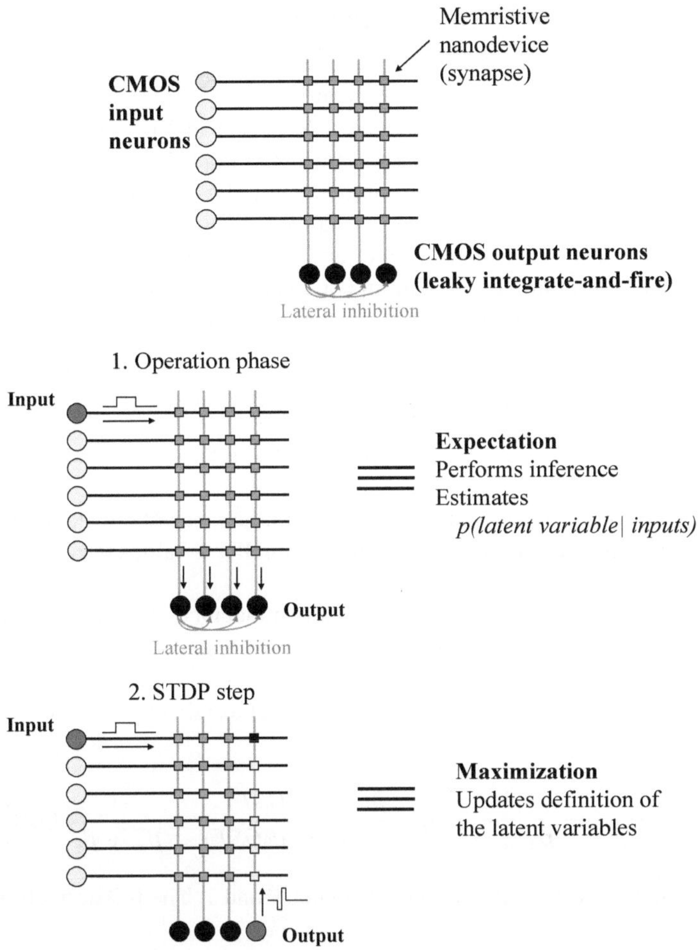

Fig. 2 *Top image* simplied schematics of the architecture considered within this chapter. *Bottom images* correspondence between the operation of this architecture and the Expectation-Maximization algorithm

3 Impact of Device Physics

Several works have shown that memristive devices can naturally implement variations of the simplified STDP rule. They are described in the other chapters of this book, and here, we revisit them in light of the theoretical analysis of the previous section.

3.1 Cumulative Memristive Synapses

We first consider the case of cumulative memristive devices. We use a simple model of the conductance increase and decrease, described in Ref. [24], which fits the measurements of a cumulative memristor presented in [12]. This model can also be employed for phase change memories associated in the 2-PCM structure [4, 30].

For the sake of simplicity, we use normalized units $w = G/G_{MAX}$, where G is the conductance of a memristive device, and G_{MAX} is its maximum conductance. We assume that minimum and maximum conductances are 0 and 1, and we identify normalized conductance with a synaptic weight w. The device model of [24] then becomes

$$\begin{aligned} \delta w_+ &= \alpha_+ \exp(-\beta_+ w) \\ \delta w_- &= \alpha_- \exp(-\beta_-(1-w)). \end{aligned} \tag{4}$$

where α_+ and α_- represent the amplitude of conductance of the memristive device changes when a programming pulse is applied. Smaller α values lead to more analog behavior.

β_+ and β_- model the dependency of this conductance change with the state of the memristive device. β values around 3.0 are typical and can, for example, model the devices of [4, 12, 30].

Under these conditions, we can show that the final weight of this particular synapse approaches

$$w_\infty = \frac{\beta_-}{\beta_+ + \beta_-} + \frac{1}{\beta_+ + \beta_-} \log \frac{p(PRE|POST)}{1 - p(PRE|POST)} + \frac{1}{\beta_+ + \beta_-} \log \frac{\alpha_+}{\alpha_-}, \tag{5}$$

with w_∞ being additionally bounded between 0 and 1. The full derivation appears in the Appendix.

In the case, where β_+ and β_- are equal ($\beta_+ = \beta_- = \beta$), Eq. (5) simplifies to

$$w_\infty = \frac{1}{2} + \frac{1}{2\beta} \log \frac{p(PRE|POST)}{1 - p(PRE|POST)} + \frac{1}{2\beta} \log \frac{\alpha_+}{\alpha_-}. \tag{6}$$

This equation is reminiscent of Eq. (3). A significant difference is that w_∞ appears to approach infinity when $p(PRE|POST)$ approaches one. However, as the weight of a physical device is bounded between 0 and 1, this divergence does not actually occur. Additionally, when considering and putting practical values into Eq. (5), it becomes similar to Eq. (3). This is shown in Fig. 3b, where Eq. (5) is plotted for different values of α_+/α_-. The value for β is taken from real devices [12]. This suggests that our inference engine with cumulative memristive devices may work by an approximation of Expectation-Maximization.

Interestingly, the curves corresponding to different values of α_+/α_- (2.0, 1.0, and 0.5) are qualitatively similar. This is in agreement with the fact that when training real

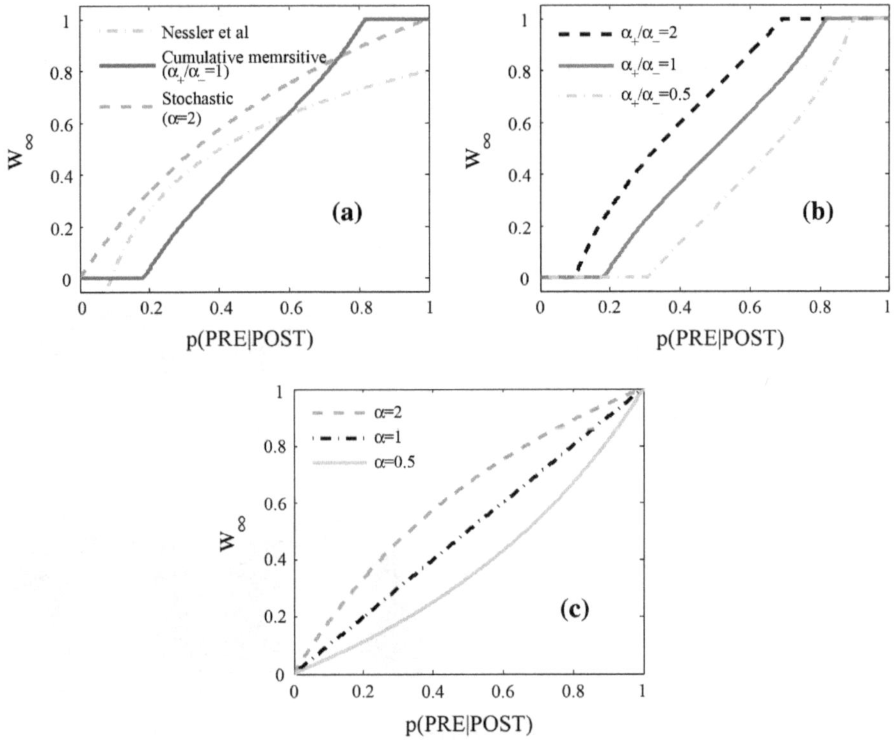

Fig. 3 **a** Final weight w_∞ as a function of $p(PRE|POST)$ Nessler theory (Eq. (3)), compared with cumulative memristive devices (Eq. (6)), and final probability of a stochastic synapse being in the state 1 (Eq. (7)). **b** Final weight w_∞ as a function of $p(PRE|POST)$ with cumulative memristive devices for different α_+ and α_- values. **c** Final probability w_∞ of a stochastic synapse being in the state 1 as a function of $p(PRE|POST)$ for different α values

life tasks with simplified STDP, the value of α_+/α_- is not a sensitive parameter. For example, on a car counting task, and with cumulative memristive devices, the best recognition rate on the four inward lanes (99 %) is obtained with $\alpha_+/\alpha_- = 2.0$. With $\alpha_+/\alpha_- = 1.0$, the recognition rate on the four inward lanes is only slightly reduced (97 %). This result has important implications when dealing with nanodevices. The parameters associated with learning do not need to be too fine-tuned for the system to be able to learn tasks.

Additionally, we notice that only the ratio α_+/α_- appears in Eq. (5) and not the actual α value. This is also consistent with what is seen when training tasks with simplified STDP. This does not mean, however, that the actual α values are entirely insignificant, as they directly affect the speed of learning.

Finally, we should note that the β value has considerable impact on the shape of the w_∞ curves. To illustrate this, we simulated a network with only one output neuron, to which we presented static photographs of faces, and repeated the simulation for devices with different β values. The resulting synaptic weights, organized as a two-

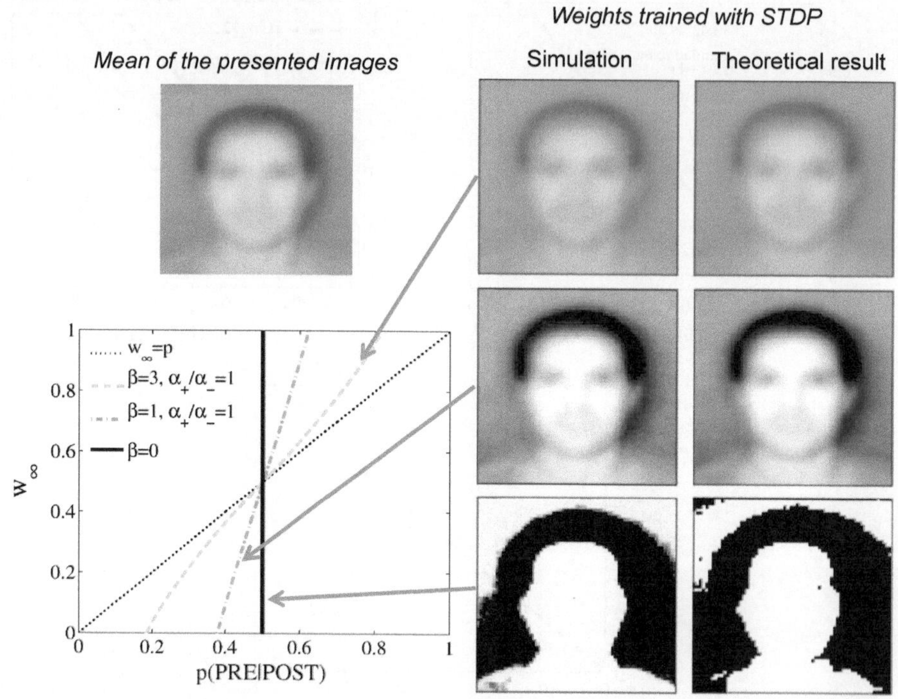

Fig. 4 *Top left picture* mean image of the faces presented in a system with STDP and one neuron, to which photographs of faces were presented. *Right pictures* representation of the final weights, as obtained in a numerical simulation and as expected from the theoretical analysis presented in this chapter. With cumulative memristive devices and *top* $\beta = 3$, $\alpha_+/\alpha_- = 1$, *middle* $\beta = 1$, $\alpha_+/\alpha_- = 1$, *bottom* $\beta = 0$. Graph: Final weight as a function of *p(PRE|POST)* in these three situations

dimensional picture, as well as the corresponding w_∞ curves, appear in Fig. 4. We can see that with a β value of 3.0 (which is close to what is observed in the devices of [12], or the 2-PCM structure [4]), the final weights are very analog and approach the mean of all the presented faces. By contrast, a β value of 1.0 produces a more binary map, amplifying the distinctive features of a face. A β value of 0 leads to an entirely binary map separating pixels where *p(PRE|POST)* is lower and greater than 0.5. This corresponds well to what would be expected from w_∞ as a function of *p(PRE|POST)* curves. This means that depending on device, different kinds of learning can thus be envisioned.

In summary, we have observed a remarkable insensitivity to relative steps of potentiation and depression, as well as to the actual value of these steps (α values). We have observed that the devices with different dependences of steps with actual values of the conductance (β values) can have different learning characteristics.

3.2 Stochastic Synapses

In some memristive devices, it is not possible to implement directly an analog of Eqs. (1) or (4), as they do not feature a cumulative analog memory behavior. By contrast, it is possible to use these devices as stochastic binary synapses: Such synapses possess only two memory states ("0" or "1"), and when a STDP step occurs, they have a *probability* to switch state. Such a *stochastic* STDP rule is illustrated in Fig. 1c. This probabilistic behavior can be implemented using pseudorandom number generators (see the examples of conductive bridge memory [31]) or intrinsic stochastic effects in nanodevices, as in spin-transfer torque magnetic tunnel junctions [32].

It has been shown that systems with binary stochastic synapses can implement the car detection task, with only one binary device per synapse, or handwritten character digit recognition using several binary devices per synapse [5, 23].

In order to understand this, we introduce p_+ the probability for a synapse to switch from low conductance ("0") to high conductance ("1") when a presynaptic spike occured before the postsynaptic spike, p_- the probability to switch from high conductance to low conductance in the other situations, and $\alpha = p_+/p_-$. At the end of the learning process, we show in the Appendix that the probability of a synapse to be in the high conductance state is

$$w_\infty = \frac{\alpha p(PRE|POST)}{1 + p(PRE|POST)(\alpha - 1).} \tag{7}$$

If p_+ and p_- are equal ($\alpha = 1$), w_∞ reduces to $p(PRE|POST)$. As shown in Fig. 3c, the shape of the w_∞ as a function of $p(PRE|POST)$ appears relatively different from that of Eq. (3), but retains some of its distinctive features. It can be thus expected that learning with stochastic synapses also performs an approximation of Expectation-Maximization in an extremely stochastic form. A more detailed analysis of this appears in Ref. [5].

When redundancy between stochastic synapses is introduced, w_∞ not only represents the probability of an individual device to be 1, but also a mean value of the weight of the equivalent synapse formed by the ensemble of the stochastic synapses. It is thus natural that the system approximates Expectation-Maximization better, as was seen with the handwritten digit classification task [5, 23].

Finally, it is insightful to compare the w_∞ curves for α values ranging from 0.5 to 2. Once again, the curves are qualitatively relatively similar. This result is consistent with the practical observation that the choice of α is not extremely sensitive when solving actual tasks with stochastic simplified STDP, although it is more sensitive than the choice of α_+/α_- in the case of cumulative memristive devices. For example, when solving the vehicle counting task with an α value of 1.0 ($p_+ = p_- = 0.1$), the detection rate is 97.3 %. With an α value of 2.0 ($p_+ = 2p_- = 0.1$), the detection rate is reduced significantly, but remains high (83.0 %). This is an important feature for being able to use a system with real devices, where mean switching probability might not be tuned with an arbitrary precision.

4 Robustness of STDP Learning

A striking feature of systems based on simplified STDP is their extreme robustness to synaptic device variations [21, 23, 32]. An understanding of the fundamental origin of this robustness is instructive. It appears to emerge from two roots: the unsupervised nature of learning and the diversity of synapses that can approximate the Expectation-Maximization algorithm.

First, the fact that the system learns in an unsupervised way is an important asset to tolerate variations. When initialized, the neurons are not specialized and respond more readily to the patterns that they are naturally capable of learning. We can, for example, consider a specific input pattern. If some synapses associated with input neurons fundamental to this pattern do not work, then the output neurons of these synapses will likely learn another input pattern. In that sense, a reasonable device variability is not deeply troublesome for the system. It may even be considered as a feature that precipitates the beginning of the learning process.

A second component of robustness to variability can be gathered from the theoretical analysis of the present chapter. For the case of cumulative devices, we have shown in Fig. 3 that the curve of w_∞ as a function of $p(PRE|POST)$ depends only on the ratio of α_+ and α_- and that its shape does not qualitatively depend dramatically on this ratio. Similarly, in the stochastic synapse case, the curve of w_∞ depends only on the ratio of p_+ and p_- and its shape does not exhibit significant qualitative dependence on this ratio. This suggests that variable synapses will still manage to perform their task even if they learn through completely different manners. This also suggests that the analysis of Sect. 3 can be an effective way to assess whether a particular technology will give rise to a robust inference engine.

5 Conclusion

In this chapter, we introduced a lead for the theoretical interpretation of the capability of systems that learn using STDP-capable memristive synapses. This theoretical analysis connects STDP learning with Bayesian inference trained with Expectation-Maximization. It can allow us to compare different device physics with regard to learning and to interpret the robustness of STDP-based learning schemes. Even more importantly, it could be a lead toward developing more advanced systems using memristive devices, capable of learning and performing complex inferences.

Acknowledgements The authors would like to thank C. Bennett, P. Bessière, L. Calvet, D. Chabi, D. Colliaux, B. De Salvo, J. Droulez, J. S. Friedman, J. Grollier, J.-O. Klein, N. Locatelli, E. Mazer, A. Mizrahi, M. Suri, S. Tiwari, D. Vodenicarevic, and W. S. Zhao. The works presented in this chapter were funded by the ANR COGNISPIN (ANR-13-JS03-0004-01) and the FP7 ICT BAMBI (FP7-ICT-2013-C) projects and by a public grant overseen by the French National Research Agency (ANR) as part of the "Investissements d'Avenir" program (Labex NanoSaclay, reference: ANR-10-LABX-0035).

Appendix Derivation of the Expressions of w_∞

Nessler et al. Synapses

We first derive Eq. (3) from Sect. 2. At the end of learning, if a synapse has reached a stable state, the impact of depression events balances the impact of potentiation events. With the notations of Sect. 2, this reads:

$$\delta w_+ p(PRE|POST) = \delta w_- p(\overline{PRE}|POST), \tag{8}$$

where we have introduced $p(\overline{PRE}|POST) = 1 - p(PRE|POST)$. By introducing the expressions of δw_+ and δw_- from Eq. (4), this becomes:

$$(C\exp(-w) - 1)p(PRE|POST) = 1 - p(PRE|POST), \tag{9}$$

which leads to Eq. (3):

$$w_\infty = \log p(PRE|POST) + \log C. \tag{10}$$

Cumulative Memristive Synapses

We now derive Eq. (5) from Sect. 3. Again, at the end of learning, if a synapse has reached a stable state, the impact of depression events balances the impact of potentiation events. With the notations of Sect. 3, this reads:

$$\delta w_+ p(PRE|POST) = \delta w_- p(\overline{PRE}|POST), \tag{11}$$

where we have introduced $p(\overline{PRE}|POST) = 1 - p(PRE|POST)$. By introducing the expressions of δw_+ and δw_- from Eq. (4), this becomes:

$$\alpha_+ \exp(-\beta_+ w_\infty) p(PRE|POST) = \alpha_- \exp(-\beta_-(1 - w_\infty))(1 - p(PRE|POST)). \tag{12}$$

Therefore, we have

$$\exp((\beta_+ + \beta_-)w_\infty - \beta_-) = \frac{\alpha_+}{\alpha_-}\frac{p(PRE|POST)}{(1 - p(PRE|POST))}, \tag{13}$$

which leads to Eq. (5):

$$w_\infty = \frac{\beta_-}{\beta_+ + \beta_-} + \frac{1}{\beta_+ + \beta_-}\log\frac{p(PRE|POST)}{1 - p(PRE|POST)} + \frac{1}{\beta_+ + \beta_-}\log\frac{\alpha_+}{\alpha_-}. \tag{14}$$

Stochastic Synapses

This appendix derives Eq. (7) from Sect. 3. Again, at the end of learning, if a synapse has reached a stable state, the impact of depression events balances the impact of potentiation events. With the notations of Sect. 3, and by introducing $p(State = 1)$ and $p(State = 0)$, the probabilities of the synapse to be in the 1 and 0 states, this reads:

$$p_+ \cdot p(PRE|POST) \cdot p(State = 0) = p_- \cdot p(\overline{PRE}|POST) \cdot p(State = 1) \qquad (15)$$

With the notations of Sect. 3, $p(State = 1) = w_\infty$ and $p(State = 0) = 1 - w_\infty$. If we introduce $\alpha = p_+/p_-$, Eq. (15) becomes

$$\alpha p(PRE|POST)(1 - w_\infty) = (1 - p(PRE|POST))w_\infty, \qquad (16)$$

which leads to Eq. (7):

$$w_\infty = \frac{\alpha p(PRE|POST)}{1 + p(PRE|POST)(\alpha - 1)}. \qquad (17)$$

References

1. Alibart, F., Pleutin, S., Bichler, O., Gamrat, C., Serrano-Gotarredona, T., Linares-Barranco, B., Vuillaume, D.: A memristive nanoparticle/organic hybrid synapstor for neuroinspired computing. Adv. Funct. Mater. **22**(3), 609–616 (2012). doi:10.1002/adfm.201101935
2. Bi, G.Q., Poo, M.M.: Synaptic modification by correlated activity: Hebb's Postulate Revisited. Annu. Rev. Neurosci. **24**(1), 139–166 (2001). doi:10.1146/annurev.neuro.24.1.139
3. Bichler, O., Querlioz, D., Thorpe, S.J., Bourgoin, J.P., Gamrat, C.: Extraction of temporally correlated features from dynamic vision sensors with spike-timing-dependent plasticity. Neural Netw. **32**, 339–348 (2012). doi:10.1016/j.neunet.2012.02.022
4. Bichler, O., Suri, M., Querlioz, D., Vuillaume, D., DeSalvo, B., Gamrat, C.: Visual pattern extraction using energy-efficient "2-PCM Synapse" neuromorphic architecture. IEEE Trans. Electron Devices **59**(8), 2206–2214 (2012). doi:10.1109/TED.2012.2197951
5. Bill, J., Legenstein, R.: A compound memristive synapse model for statistical learning through STDP in spiking neural networks. Front. Neurosci. **8**, 412 (2014). doi:10.3389/fnins.2014.00412
6. Chabi, D., Querlioz, D., Zhao, W., Klein, J.O.: Robust learning approach for neuro-inspired nanoscale crossbar architecture. J. Emerg. Technol. Comput. Syst. **10**(1), 5:1–5:20 (2014). doi:10.1145/2539123
7. Chanthbouala, A., Garcia, V., Cherifi, R.O., Bouzehouane, K., Fusil, S., Moya, X., Xavier, S., Yamada, H., Deranlot, C., Mathur, N.D., Bibes, M., Barthélémy, A., Grollier, J.: A ferroelectric memristor. Nat. Mater. **11**(10), 860–864 (2012). doi:10.1038/nmat3415
8. Dempster, A., Laird, N., Rubin, D.: Maximum likelihood from incomplete data via the EM algorithm. J. Roy. Stat. Soc. Ser. B (Methodological) **39**(1), 1–38 (1977). doi:10.2307/2984875
9. Erokhin, V., Berzina, T., Camorani, P., Smerieri, A., Vavoulis, D., Feng, J., Fontana, M.P.: Material memristive device circuits with synaptic plasticity: learning and memory. BioNanoSci. **1**(1–2), 24–30 (2011). doi:10.1007/s12668-011-0004-7

10. Gacem, K., Retrouvey, J.M., Chabi, D., Filoramo, A., Zhao, W., Klein, J.O., Derycke, V.: Neuromorphic function learning with carbon nanotube based synapses. Nanotechnology **24**(38), 384013 (2013). doi:10.1088/0957-4484/24/38/384013
11. Indiveri, G., Linares-Barranco, B., Legenstein, R., Deligeorgis, G., Prodromakis, T.: Integration of nanoscale memristor synapses in neuromorphic computing architectures. Nanotechnology **24**(38), 384010 (2013). doi:10.1088/0957-4484/24/38/384010
12. Jo, S.H., Chang, T., Ebong, I., Bhadviya, B.B., Mazumder, P., Lu, W.: Nanoscale memristor device as synapse in neuromorphic systems. Nano Lett. **10**(4), 1297–1301 (2010). doi:10.1021/nl904092h
13. Lecerf, G., Tomas, J., Saighi, S.: Excitatory and inhibitory memristive synapses for spiking neural networks. In: 2013 IEEE International Symposium on Circuits and Systems (ISCAS), pp. 1616–1619 (2013). doi:10.1109/ISCAS.2013.6572171
14. Lee, J.H., Likharev, K.K.: Defect-tolerant nanoelectronic pattern classifiers. Int. J. Circuit Theor. Appl. **35**(3), 239–264 (2007). doi:10.1002/cta.410
15. Linares-Barranco, B., Serrano-Gotarredona, T.: Exploiting memristance in adaptive asynchronous spiking neuromorphic nanotechnology systems. In: Proceedings of IEEE Conference on Nanotechnology, 2009, pp. 601–604 (2009)
16. Markram, H., Lubke, J., Frotscher, M., Sakmann, B.: Regulation of synaptic efficacy by coincidence of postsynaptic APs and EPSPs. Science **275**(5297), 213–215 (1997). doi:10.1126/science.275.5297.213
17. Masquelier, T., Guyonneau, R., Thorpe, S.J.: Spike timing dependent plasticity finds the start of repeating patterns in continuous spike trains. PLoS ONE **3**(1), e1377 (2008). doi:10.1371/journal.pone.0001377
18. Masquelier, T., Thorpe, S.J.: Unsupervised learning of visual features through spike timing dependent plasticity. PLoS Comput. Biol. **3**(2), e31 (2007). doi:10.1371/journal.pcbi.0030031
19. Nessler, B., Pfeiffer, M., Buesing, L., Maass, W.: Bayesian computation emerges in generic cortical microcircuits through spike-timing-dependent plasticity. PLoS Comput. Biol. **9**(4) (2013). doi:10.1371/journal.pcbi.1003037
20. Pershin, Y.V., La Fontaine, S., Di Ventra, M.: Memristive model of amoeba learning. Phys. Rev. E **80**(2), 021926 (2009). doi:10.1103/PhysRevE.80.021926
21. Querlioz, D., Bichler, O., Dollfus, P., Gamrat, C.: Immunity to device variations in a spiking neural network with memristive nanodevices. IEEE Trans. Nanotechnol. **12**(3), 288–295 (2013)
22. Querlioz, D., Bichler, O., Gamrat, C.: Simulation of a memristor-based spiking neural network immune to device variations. In: Proceedings of the International Joint Conference on Neural Networks (IJCNN), pp. 1775–1781 (2011)
23. Querlioz, D., Bichler, O., Vincent, A., Gamrat, C.: Bioinspired programming of memory devices for implementing an inference engine. Proc. IEEE **103**(8), 1398–1416 (2015). doi:10.1109/JPROC.2015.2437616
24. Querlioz, D., Dollfus, P., Bichler, O., Gamrat, C.: Learning with memristive devices: how should we model their behavior? In: Proceedings of IEEE/ACM International Symposium Nanoscale Architectures (NANOARCH 2011), p. 150 (2011)
25. Seo, K., Kim, I., Jung, S., Jo, M., Park, S., Park, J., Shin, J., Biju, K.P., Kong, J., Lee, K., Lee, B., Hwang, H.: Analog memory and spike-timing-dependent plasticity characteristics of a nanoscale titanium oxide bilayer resistive switching device. Nanotechnology **22**(25), 254023 (2011). doi:10.1088/0957-4484/22/25/254023
26. Sharad, M., Augustine, C., Panagopoulos, G., Roy, K.: Spin-based neuron model with domain-wall magnets as synapse. IEEE Trans. Nanotechnol. **11**(4), 843–853 (2012). doi:10.1109/TNANO.2012.2202125
27. Snider, G.: Spike-timing-dependent learning in memristive nanodevices. In: Proceedings of IEEE International Symposium on Nanoscale Architectures 2008 (NANOARCH), pp. 85–92 (2008). doi:10.1109/NANOARCH.2008.4585796
28. Snider, G.S.: Self-organized computation with unreliable, memristive nanodevices. Nanotechnology **18**(36), 365202 (2007). doi:10.1088/0957-4484/18/36/365202

29. Strukov, D.B., Snider, G.S., Stewart, D.R., Williams, R.S.: The missing memristor found. Nature **453**(7191), 80–83 (2008). doi:10.1038/nature06932
30. Suri, M., Bichler, O., Querlioz, D., Traoré, B., Cueto, O., Perniola, L., Sousa, V., Vuillaume, D., Gamrat, C., DeSalvo, B.: Physical aspects of low power synapses based on phase change memory devices. J. Appl. Phys. **112**(5), 054904–054904–10 (2012). doi:10.1063/1.4749411
31. Suri, M., Querlioz, D., Bichler, O., Palma, G., Vianello, E., Vuillaume, D., Gamrat, C., DeSalvo, B.: Bio-inspired stochastic computing using binary CBRAM synapses. IEEE Trans. Electron Devices **60**(7), 2402–2409 (2013). doi:10.1109/TED.2013.2263000
32. Vincent, A., Larroque, J., Locatelli, N., Ben Romdhane, N., Bichler, O., Gamrat, C., Zhao, W., Klein, J.O., Galdin-Retailleau, S., Querlioz, D.: Spin-transfer torque magnetic memory as a stochastic memristive synapse for neuromorphic systems. IEEE Trans. Biomed. Circuits Syst. **9**(2), 166–174 (2015). doi:10.1109/TBCAS.2015.2414423
33. Yu, S., Wu, Y., Jeyasingh, R., Kuzum, D., Wong, H.P.: An electronic synapse device based on metal oxide resistive switching memory for neuromorphic computation. IEEE Trans. Electron Devices **58**(8), 2729–2737 (2011). doi:10.1109/TED.2011.2147791

Erratum to: Novel Biomimetic Si Devices for Neuromorphic Computing Architecture

U. Ganguly and Bipin Rajendran

Erratum to:
Chapter "Novel Biomimetic Si Devices for Neuromorphic Computing Architecture" in: M. Suri (ed.), *Advances in Neuromorphic Hardware Exploiting Emerging Nanoscale Devices*, Cognitive Systems Monographs 31, DOI 10.1007/978-81-322-3703-7_8

The original version of this Chapter "Novel Biomimetic Si Devices for Neuromorphic Computing Architecture" was inadvertently published with an incorrect affiliation for the author Bipin Rajendran. The correct information is given below: Dr. Bipin Rajendran Department of Electrical and Computer Engineering, New Jersey Institute of Technology, Newark, NJ, USA

The updated original online version for this chapter can be found at DOI 10.1007/978-81-322-3703-7_8

U. Ganguly (✉)
Department of Electrical Engineering, IIT Bombay, Mumbai 400076, India
e-mail: udayan@ee.iitb.ac.in

B. Rajendran
Department of Electrical and Computer Engineering,
New Jersey Institute of Technology, Newark, NJ, USA

© Springer (India) Pvt. Ltd. 2017

M. Suri (ed.), *Advances in Neuromorphic Hardware Exploiting Emerging Nanoscale Devices*, Cognitive Systems Monographs 31,
DOI 10.1007/978-81-322-3703-7_11